Biodiversity Challenges
– *A Way Forward* –

THE EDITOR

Dr. Shashikanth H. Majagi did his M. Sc. and Ph. D. in Zoology from Gulbarga University, Kalaburagi, Karnataka. He started his carrier as Assistant professor in Government First Grade College, Chikkballapur and Government College (Autonomous) Kalaburagi. He has published 45 research papers and book articles. He has attended and presented 50 research papers in national/international conferences/symposia and 06 books to his credit. He is a reviewer/editorial board member for many journals of international acclaim. He is the life member for many academic bodies. He has given radio talks and Doordarshan programmes as well . Received prestigious Award for Research Paper (ARP) award from VGST, Government of Karnataka and IOSRD Life time award and Novel Research Academy Award of Excellence in Research. He has successfully organized two national Conference and Workshops. Successfully completed research projects funded by is UGC. Successfully Guided 02 Ph. D. students and 03 are working. He has worked as Special officer for NAAC, Department of Collegiate Education, Regional Joint Director Office, Kalaburagi. Currently he is teaching Environmental biology, Biodiversity and Limnology to Master degree Students. His research interests are on Limnology, Water quality studies and biodiversity. Currently working as Associate Professor and Chairman, Department of Zoology, Deputy Registrar (DPAR) and Nodal officer for Youth Red Cross.

Biodiversity Challenges
— A Way Forward —

– Editor –

Shashikanth Majagi

Associate Professor and Chairman
Department of Studies in Zoology,
Vijaynagar Sri Krishnadevaraya University,
Ballari – 583105, Karnataka

2022

Daya Publishing House®
A Division of
Astral International Pvt. Ltd.
New Delhi – 110 002

ISBN: 9789354614019

Publisher's Note:

Every possible effort has been made to ensure that the information contained in this book is accurate at the time of going to press, and the publisher and author cannot accept responsibility for any errors or omissions, however caused. No responsibility for loss or damage occasioned to any person acting, or refraining from action, as a result of the material in this publication can be accepted by the editor, the publisher or the author. The Publisher is not associated with any product or vendor mentioned in the book. The contents of this work are intended to further general scientific research, understanding and discussion only. Readers should consult with a specialist where appropriate.

Every effort has been made to trace the owners of copyright material used in this book, if any. The author and the publisher will be grateful for any omission brought to their notice for Acknowledgment in the future editions of the book.

Published by : **Daya Publishing House®**
A Division of
Astral International Pvt. Ltd.
– ISO 9001:2015 Certified Company –
4736-61/23, Ansari Road, Darya Ganj
New Delhi-110 002
Ph. 011-43549197, 8130496929
e-mail: info@astralint.com
Website: www.astralint.com

— *Dedicated to* —

My Beloved Sister

Late Smt. Manjula R.

Acknowledgements

I wish to express my deep sense of gratitude to all the research paper contributors who were very generous for sending their original book chapters.

I am thankful to Dr. Vijaykumar Malashetty, Professor, Department of Zoology, Vijayanagar Sri Krishnadevaraya University, Ballari for writing foreword to this book.

I Profusely thank Dr. K. Vijaykumar, Professor and Chairman, Department of Zoology, Gulbarga University, Kalaburagi and Dr. B. B. Hosetti Professor (Retd.) Kuvempu University, Shankarghatta, for reviewing the research papers.

My sincere thanks to Doddappa Sri Manikappa Majagi and Doddamma Smt. Sukumari and my parents Sri. Hanmantappa Majagi and Shakuntala Laxmibai, Wife Dr. Anita S. Majagi and daughters Miss. Sinchana and Miss. Saanidhya.

Shashikanth Majagi

Foreword

The book before you offers an overall view of this biological diversity and carries the urgent warning that we are rapidly altering and destroying the environment, its challenges and how to way forward to protect. Thus it gives me a great pleasure to contribute this foreword.

As I read the book prior to writing this Foreword, I was impressed by many unique features that I would like to share with you. Biodiversity is life! The diversity of life forms, so numerous that we have yet to identify most of them, is the greatest wonder of this planet. The biosphere is an intricate tapestry of interwoven life forms. The major threats to biodiversity that result from human activity are habitat loss, fragmentation and degradation, over exploitation of species for human use, introduction of exotic species and possibility of spread of disease. This leads to threats to the threatened species and result in the extinction of such species. Typically these threats develop so rapidly and on such a large scale that species are not able to adopt genetically to the changes or disperse to more hospitable location. The rapid and visible changes in environmental variables within few weeks of the lockdown due to COVID-19 were surprising even for experts which should create an optimistic attitude towards biodiversity conservation. The scientific community has to lead from the front, in creating solutions and in steering the socio-political will required to implement these solutions for a more long lasting process of environmental conservation. Besides there is an urgent need to create awareness among rural and urban population to highlight the importance of conserving biodiversity.

Today, the issues and challenges relating to wildlife and biodiversity conservation are embedded in understanding the human dimension with its social, cultural, political, economic and legal complexities. An interdisciplinary approach to challenges like that of human–wildlife conflict, will help scientists to arrive at better solutions that might ensure conservation of nature in the longer run.

Biodiversity Challenges – A Way Forward is a edited book, Many scientists have generously contributed in this endeavour, an updating of many of the principal issues in conservation biology and resource management. It also documents a new alliance between scientific, governmental, and commercial forces—one that can be expected to reshape the international conservation movement for decades to come. This book also accumulated enough data on deforestation, species extinction, and tropical biology to bring global problems into sharper focus and warrant broader public exposure. It also a handy resource for the local conservationist community consisting of policy makers, academicians, scientists, environmentalists, students and gross root level conservationists. It also provides an invaluable toolkit for a large and under-resourced audience of students in developing nations and addresses the key issues in conservation biology, clearly stating the challenges but also offering solutions.

Dr. Shashikanth H. Majagi is currently Associate Professor of Zoology at Vijayanagara Sri Krishnadevaraya University, Ballari and one of the leading Environmental Biologist of the country. "Dream is not what you see while sleeping. Dream is something that doesn't let you sleep" – rightly quoted by Dr. A.P.J. Abdul Kalam. His broad research interests in the field of biodiversity, strong determination and dedication made him to bought this edited book. I know that you will soon have a much larger audience for your work.

Congratulations!

Dr. Vijaykumar B. Malashetty, Ph.D

Professor,

Department of Studies in Zoology,

Vijayanagara Sri Krishnadevaraya University, Ballari

Preface

The present book provides an information about biodiversity, how its rise, about its distribution, importance and how to maintain it, representing its point of view into the literature on biological diversity. Biodiversity continues to decline, and the main drivers of biodiversity loss continue to increase. The unsustainable use of natural resources has led to an alarming rate of habitat loss. Climate change, invasive species, and pollution pose additional threats to global species richness. The present book neither complies of a technical research reports or the complete action plans and nor the detailed documentation on biodiversity importance or the issues for the loss of biodiversity. Instead, various recent research articles, here represents the relevant ideas and goals in the critical importance to sustainable development and a conservation measures on biodiversity. The biodiversity loss or crisis is one of the major global concerns and every country or community has vital role in the challenges ahead. The chapters in the book has its own impact on values, distributions and maintenance of biodiversity as of which we tried to draw out the major problems and reveal the ground reports with actual examples. This book might form a relevant source for the readers and hope to be a useful resource in exploring the main stream themes at greater levels and details.

Shashikanth Majagi

Contents

Chapter 1

Cropping Systems: A Comparative Account

Naveena Nazim, Mushtaq Rasool Mir and Mehreen Manzoor*

College of Temperate Sericulture, Mirgund, SKUAST-Kashmir, Post Box No.674, GPO-Srinagar, J&K

**Correspondending author e-mail: naveenanazim08@gmail.com*

"Cropping system" is a critical aspect in developing an effective ecological farming system to manage and organize crops so that they best utilize the available resources (soil, air, sunlight, water, labour, equipments). It represents cropping patterns used on a farm and their interaction with farm resources, other farm enterprises and available technology which determine their makeup. It is executed at field level. Cropping systems based on climate, soil and water availability have to be evolved for realizing the potential production levels through efficient use of available resources. The cropping system should provide enough food for the family, fodder for cattle and generate sufficient cash income for domestic and cultivation expenses.

Keywords: *Resources, Climate, Food, Domestic.*

Introduction

Cropping system comprises all components required for the production of a particular crop and the interrelationships between them and environment. In the cropping systems, sometimes a number of crops are grown together or they are grown separately at short intervals in the same field so that the available resources (soil, air, sunlight, water, labour, equipments *etc.*) are utilized to the best. Andrew and Kassam (1976) defined cropping system as- cropping patterns used on a farm and their interaction with farm resources and farm enterprises and available technology which determine their makeup. Singh *et al.* (1986) generalized the cropping system as "the principles and practices of cropping and their interaction with farm resources, technology, aerial and edaphic environment to suit the regional or national or global needs and production strategy. Sebillotte, (1990), considered cropping system as a set of management procedures applied to a given, uniformly treated agricultural areas; this can be a field, part of a field, or a group of fields. The cropping pattern change is of great interest to the agricultural economists for its notable impact on agricultural output (Ranade, 1980). The cropping pattern usually changes over time with the development of agriculture, as is evident in the case of agriculture in India (Vyas, 1996). This type of change is largely characterized by an increasing trend towards commercial crops over the years (Nadkarni and Vedini, 1996). According to Francis *et al.* (1976), Willey (1979), Smith and Francis (1986) the significance of genotype by cropping system interaction is the crucial point when evaluating different cultivars for their performance under different cropping systems to decide whether separate selection programs should be applied under different cropping systems such as monoculture or intercropping. When land is limited, intensive cropping is adopted to fully utilize available water and labour. When sufficient and cheap labour is available, vegetable crops are also included in the cropping system. When capital is not a constraint, commercial crops like sugarcane, banana, turmeric *etc.* can be fitted in cropping system.

Basic Principles of Cropping System

- Crops selected have a complementary influence on each other.
- Efficient utilization of resources.
- Enhancement in soil fertility.
- Diversity of species.

Objectives of Cropping Systems

- Efficient utilization of all resources *viz.* land, water, and solar radiation maintaining stability in production and obtaining higher net returns. (The efficiency is measured by the quantity of produce obtained per unit resource in a unit time).
- Maintenance of soil fertility and crop growth.
- Control of pests, diseases and weeds.
- Crop insurance.
- Food and financial security.

Classifications of Cropping System

Depending on the resources and technology available cropping systems are broadly classified into two types *viz.*- mono-cropping and multiple- cropping (Arun 2017), which in turn are divided into different types:

1. Monocropping (Sole cropping)
2. Multiple cropping

1. Monocropping

It is the agricultural practice of producing or growing a single crop, plant variety or breed in a field or farming system at a time. It is also known as solid planting. Palaniappan and Sivaraman (1996), defined monocropping as growing only one crop on a piece of land year after year.

Advantages

- ☆ There is less competition for sunlight, nutrients and space.
- ☆ The crop is same over the whole farm, therefore farm management decisions are simplified.
- ☆ Ease of irrigation, weed management and harvest.
- ☆ Controlling the pests and disease becomes relatively easy.
- ☆ One crop means one set of harvest, planting and cultivation equipment.

Disadvantages

- ☆ High risk of total loss in case of crop failure.
- ☆ It is difficult to maintain cover on the soil.
- ☆ It encourages crop specific pests, diseases and weeds.
- ☆ It can reduce the soil fertility and change the soil structure.

Monocropping

2. Multiple Cropping

Growing two or more crops on the same piece of land in a year instead of one crop (Palaniappan and Sivaraman 1996). It is the intensification of cropping in time and space dimensions *i.e* more number of crops within a year and more number of crops on the same piece of land at any given period. The principal limiting factors in adoption of multiple cropping are low intensity of irrigation and preponderance of long duration varieties. Multiple cropping has been practiced for centuries by small-scale farmers to reduce the risk of crop failure, attain higher yields, and to improve soil fertility (Litsinger and Moody, 1976). Traditional multiple cropping systems are estimated to still provide as much as 15-20 per cent of the world's food supply (Altieri, 1999).

Types of Multiple Cropping

a) Crop rotation/Sequential cropping

b) Ratoon cropping/stubble cultivation

c) Intercropping

a) Crop Rotation/Sequential Cropping

Growing two or more crops in sequence on the same field in a farming year. The succeeding crop is planted after the preceding crop has been harvested. It is of following types:

- ☆ **Double cropping:** Growing of two crops in a year in sequence. *e.g* : paddy – mustard/oats, green gram- jowar
- ☆ **Triple cropping:** Growing of three crops in a year in sequence. *e.g*: rice-potato-groundnut, Cowpea- mustard- jute
- ☆ **Quadruple cropping:** Growing of four crops in a year in sequence. *e.g* : maize – toria – potato – wheat green gram - maize – toria – wheat

Crop Rotation

Principles of Crop Rotation

- ☆ Crops with tap roots should be followed by crops with a fibrous root system as this helps in the proper and uniform use of nutrients from the soil.
- ☆ Legumes should be grown after non legumes as they fix atmospheric nitrogen into the soil and add more organic matter to the soil.
- ☆ Exhaustive crops (crops which need more inputs like more fertilizer, irrigation, insecticide *etc.*) should be followed by less exhaustive crops, which require less care (*i.e.*, potato should be followed by leguminous crops).
- ☆ Selection of crop should be demand-based (that is, crops needed by the market should be chosen as it can be sold at a higher price).
- ☆ Crop selection also depends on land type, irrigation facilities, soil and climatic considerations. Financial constraints of the farmer should also be kept in mind.

Advantages of Crop Rotation

In addition to increased crop yields and profit, the following are the advantages of crop rotation over monoculture.

1. Crop rotation helps to create biodiverse soils which reduce the chances of soil bound organisms to get a pest and reducing the use of pesticides.
2. Crop rotation helps to minimize the growth of herbs getting a weed that needs to be eradicated and so reducing the use of pesticides.
3. Crop rotation helps getting a good soil structure (alternatively shallow and deep rooting roots), high organic matter, good water provision, especially in combination with conservation tillage, resulting in a higher yield.
4. Crop rotation, especially performed with nitrogen-fixating rotation crops, will reduce the input of fertilizers and so reduce the pollution by nitrogen.
5. Crop rotation, especially combined with conservation tillage, will lead to higher soil-carbon content and so contribute to combating climate change.

b) Ratoon Cropping/Stubble Cultivation

One of the important methods of intensive cropping, allowing the stubbles of the original crop to strike again after harvesting and to raise another crop. Harvesting leaves behind plenty of opportunities for latent basal buds to develop again.

Sugarcane

Fennel

Ratoon Cropping

c) *Intercropping*

Intercropping is defined as the cultivation of two or more crop species in the same field for the whole or part of their growing period (Willey, 1990; Hauggaard-Nielsen *et al.,* 2008). According to Cardinale *et al.,* 2007- Intercropping is a practical application of the principle of productivity increase by biodiversity. Andrews and Kassam, 1976; Ofori and Stern, 1987; Anil *et al.,* 1998, defined intercropping as the agricultural practice of cultivating two or more crops in the same space at the same time. It differs from crop rotation in which two or more crops are grown one after the other. Many studies have indicated that intercropping is advantageous over sole cropping, having benefits such as higher land and/or resource use efficiency (Reddy and Willey, 1981; Vandermeer, 1992; Zhang and Li, 2003; Lithourgidis *et al.,* 2011) and could achieve food security with reduced anthropogenic inputs and lower environmental impacts (Brooker *et al.,* 2015). Intercropping can provide ecosystem services beyond food production including the suppression of pests and diseases (Andow, 1991; Trenbath, 1993), the suppression of weeds (Ayeni *et al.,* 1984; Liebman and Dyck, 1993; Banik *et al.,* 2006), and an increase in the retention of organic soil carbon and nitrogen (Cong *et al.,* 2015), which in turn leads to a lower need for chemical inputs and the associated environmental impacts compared to sole crops (Lithourgidis *et al.,* 2011). Intercropping as an agricultural strategy could therefore contribute to sustainable intensification of agriculture (Tilman *et al.,* 2011; Bedoussac *et al.,* 2015; Jensen *et al.,* 2015) and is able to produce more food on limited arable land while exerting less impact on the environment than sole crops due to better exploitation of ecological mechanisms (Bedoussac *et al.,* 2015; Jensen *et al.,* 2015**)**. Before the 1940s, intercropping was commonly practiced in the United States and Europe (Andersen, 2005; Machado, 2009). It however has gradually disappeared in developed countries due to mechanization and the availability of cheap synthetic fertilizers and pesticides which make sole cropping an efficient way to go (Horwith, 1985; Machado, 2009). In developing countries like- Asia, Africa and Latin America, where farmers have limited access to mechanization and agricultural chemicals, intercropping is still widely used (Machado, 2009; Lithourgidis *et al.,* 2011). A number of research workers have found that crop combinations having shade tolerant legumes with non-climbing habit such as groundnut, cowpeas, soyabean *etc.* with maize, sorghum, millet, cotton *etc.* have given greater overall yield from intercropping compared to sole cropping (Bodade, 1964; Enyi, 1973). One important reason for which intercropping is popular in the developing world is that it is more stable than monocropping (Willey and Reddy, 1981; Horwith, 1985; Ofori and Stern, 1987). For hundreds of years, many crops have been grown in association with one another and crop mixtures probably represent some of the first farming systems practiced (Plucknett and Smith, 1986). The general trend in most intercropping experiments is that the yield of a given crop in the mixture is less than the yield of the same crop grown alone, but the total productivity per unit of land is usually greater for mixtures than for sole crops (Wiley, 1979; Zhang and Li, 2003; Li *et al.,* 2013). According to Clawson, 1985., If one crop of the mixture fails due to adverse environmental conditions, the other crop(s) might withstand the stresses and remain harvestable and profitable, in other words a Single crop may often fail because of adverse conditions such as frost, drought, flood, or even pest attack, so farmers reduce their risk for total crop failure by growing more than one crop in their field.

There are some socio economic (Ofori and Stern, 1987), biological and ecological advantages (Aggarwal *et al.,* 1992; Fininsa, 1996) in intercropping over monocropping *e.g.,* intercropping with legumes is an excellent practice for controlling soil erosion and sustaining crop production (El-Swaify *et al.,* 1988), in areas where there is water scarcity, intercropping is suitable methods (Lynam *et al.,* 1986), increased leaf cover in intercropping systems helps to reduce weed populations once the crops are established (Mugabe *et al.,* 1982; Beets, 1990), Singh and Adjeigbe (2002) stated that monocropping needs more chemical to control pests and diseases than intercropping, intercropping often provides higher cash return than growing one crop alone (Ahmad and Rao, 1982; Grimes *et al.,* 1983; Kurata, 1986).

The main advantage of intercropping is the more efficient utilization of the available resources and the increased productivity compared with each sole crop of the mixture (Willey, 1979; Jannasch and Martin, 1999; Li *et al.,* 1999; Hauggaard-Nielsen *et al.,* 2001; Zhang and Li, 2003; Szumigalski and Van Acker, 2006; Andersen *et al.,* 2007; Dhima *et al.,* 2007; Ofosu-Anim and Limbani, 2007; Muoneke *et al.,* 2007; Agegnehu *et al.,* 2008; Carrubba *et al.,* 2008; Launay *et al.,* 2009; Mucheru-Muna *et al.,* 2010). Research has shown that intercrops are most productive when component crops differ greatly in growth duration (Wien and Smithson, 1981; Smith and Francis, 1986; Fukai and Trenbath, 1993; Keating and Carberry, 1993).

Intercropping has been an important production practice in many parts of the world and it continues to be an important farming practice in developing countries (Clawson, 1985).

Principles of Intercropping

- ☆ The crops grown in association should have complementary effects rather competitive effects.
- ☆ The subsidiary crop should be of shorter duration and of faster growing habits to utilize the early slow growing period of main crop and they must be harvested when main crop starts growing.
- ☆ The planting method and management should be simple, less time consuming, economical and profitable so that it may have wider adoptability.
- ☆ Component crops of similar pests and disease pathogens and parasite infestations should not be chosen.
- ☆ There are at least six types of intercropping according to spatial arrangement:
  - (i) Row intercropping:
  - (ii) Strip intercropping:
  - (iii) Mixed intercropping:
  - (iv) Relay intercropping:
  - (v) Alley Cropping:
  - (vi) Multistoried Cropping/Multi-tier Cropping:

i) Row Intercropping

Row intercropping is the growing of two or more crops simultaneously at the same time with at least one crop planted in rows. *e.g.* maize + greengram, maize + blackgram, groundnut + redgram. This strategy is an efficient way of maximizing the use of farm land by utilizing vacant spaces while at the same time suppressing the growth of weeds during the juvenile stage of the main crop.

Row-intercropping

ii) Strip Intercropping

Strip intercropping is the growing of two or more crops simultaneously in different strips wide enough to permit independent cultivation but narrow enough for the crops to interact agronomically. The most common crop choices for strip cropping are closely sown crops such as wheat, or other forages which are alternated with strips of row crops, such as corn, soya beans, cotton, or sugar beets.

Strip Intercropping

iii) Mixed Intercropping

Mixed intercropping or mixed cropping is the growing of two or more crops at the same time with no distinct row arrangement. This system of cropping is practised in areas where climatic hazards such as flood, drought, frost *etc.* are frequent and common. The farmers always fear that their crops will fail. Mixed cropping is also practised with a view to achieve multiple requirements of food and fibre. Under mixed cropping, the time of sowing of all the crops is almost the same. However, they may mature either together (wheat + gram, wheat + barley or wheat + mustard) or they may mature at different times (arhar + Jowar, til + mung or bajra + groundnut).

Based on the method of sowing mixed cropping can be classified into the following groups:

1. **Mixed Crops** – In this case, the seeds of different crops are mixed together and then sown either in lines or they are broadcasted. The system is not scientific and it causes problems in performing all the agricultural operations and harvesting of the crops. On the other hand some researchers like -Altieri and Liebman, 1986; Zuofa *et al.,* 1992 reported that mixed cropping reduces weed incidence.

Mixed Intercropping

2. **Guard crops** – Under this system of cropping, the main crop is grown in the centre, surrounding by hardy or thorny crops such as safflower around pea or wheat, mesta (patsan) around sugarcane, jowar around maize *etc.* with a view to provide protection to the main crop.

Guard Crops

3. **Augmenting Crops** – When sub crops are sown to supplement the yield of the main crop, the sub crops are called as augmenting crops such as Japanese mustard with berseem. Here the mustard helps in getting higher tonnage of fodder in spite of the fact that berseem gives poor yield in first cutting.

Augmenting Crops

iv) Relay Intercropping/Over Lapping Cropping

Relay intercropping or relay cropping is a system of growing two or more crops simultaneously during part of the life cycle of each. A second crop is planted after the first crop has flowered (reached its reproductive stage) but before it is ready for harvest. There is thus a minimum temporal overlap of two or more crops. This practice requires some kind of temporal separation, *e.g.* different planting dates of the component crops so that the differential influence of weather and in particular temperature on component crop growth can be modified (Midmore, 1993). Such relay intercropping could better exploit the length of the growing season (Lithourgidis *et al.*, 2011) and intercept more light over the whole growing season (Keating and Carberry, 1993; Zhang *et al.*, 2008) than sole crops. According to Enyi, 1977, crops which mature at different times thereby separating their periods of maximum demand to nutrient and moisture aerial space and light could be suitably intercropped. Examples of relay crops are sweet potato with corn, red gram/chickpea with wheat, cowpea with bajra, green gram with jowar *etc.*

Relay Intercropping

v) Alley Cropping

Alley cropping is broadly defined as the planting of two or more sets of single or multiple rows of trees or shrubs at wide spacing, creating alley ways within which agricultural, horticultural, or forage crops are cultivated. Alley cropping can diversify farm income, improve crop production and provide protection and conservation benefits to crops. Common examples of alley cropping plantings include:

- ☆ Wheat, corn or soya beans planted in between rows of walnut or almond trees.
- ☆ Forage crops can be cultivated between tree rows for harvest or livestock grazing.

Alley Cropping

vi Multistoried Cropping or Multi-tier Cropping

Growing plants of different height in the same field at the same time is termed as multistoried cropping. It is mostly practised in orchards and plantation crops for maximum use of solar energy even under high planting density.

Examples

A) Eucalyptus + Papaya + Berseem.

B) Coconut +Pineapple +Turmeric/Ginger.

C) Sometimes it is practiced under field crops such as

- ☆ Sugarcane+ Potato + Onion
- ☆ Sugarcane + Mustard + Potato

Multistoried Cropping

Advantages of Intercropping

- ☆ Reduces the insect/mite pest populations because of the diversity of the crops grown. When other crops are present in the field, the insect/mite pests are confused and they need more time to look for their favorite plants.

- Reduces the plant diseases. The distance between plants of the same species is increased because other crops (belonging to a different family group) are planted in between.
- Attracts more beneficial insects, especially when flowering crops are included the cropping system.
- Minimizes labour cost on the control of weeds. A mixture of various crops gives often a better coverage of the soil leaving less space for the development of weeds.
- Utilizes the farm area more efficiently.
- Results in potential increase for total production and farm profitability than when the same crops are grown separately.
- Provides two or more different food crops for the farm family in one cropping season.
- Insurance against total crop failure under aberrant conditions or pest/disease epidemics.
- Reduction of water evaporation from the soil surface.
- Controls soil erosion by preventing rain drops from hitting the bare soil.

Disadvantages

- Depending on crops mixed, competition for light, water and nutrients, or allelopathic effects that may occur between mixed crops may reduce yields (Cenpukdee and Fukai, 1992a, 1992b; Carruthers *et al.*, 2000; Santalla *et al.*, 2001; Yadav and Yadav, 2001; Olowe and Adeyemo, 2009).
- Harvesting of one crop component may cause damage to the others.
- Higher relative humidity in the air can favour disease outbreak, especially of fungi.
- Leaf, branch, fruit, or water-drop fall from taller elements in a mixed crop system can damage shorter ones.
- The systems are more complex and less understood agronomically and biologically. Statistical designs for experiment analysis are much more complex.
- These systems require more hand labour, which can be considered a disadvantage in some circumstances.
- For producers with limited economic resources, it may take longer to recover the entire initial investment.
- Possible proliferation of harmful animals (especially rodents and insects).

REFERENCES

Agegnehu G, Ghizaw A, Sinebo W. (2008). Yield potential and land-use efficiency of wheat and faba bean mixed intercropping. Agron Sustain Dev 28: 257-263.

Aggarwal, P.K., D.P. Garrity, S.P. Liboon and R.A. Morris, (1992). Resource use and plant interactions in a rice mungbean intercrop. Agron. J., 84: 71-78.

Ahmad, S. and M.R. Rao, (1982). Performance of maize-soybean intercrop combination in the tropics: Results of a multilocation study. Field Crops Res., 5: 147-161.

Altieri M.A. (1999). The ecological role of biodiversity in agro-ecosystems. Agr Ecosyst Environ 74: 19-31.

Altieri, M.A. and M. Liebman, (1986). Insect Weed and Plant Disease Management in Multiple Cropping Systems. In: Multiple Cropping Systems, Francis, C.A. (Ed.). Mac Millan Publ.Co., NewYork.

Andersen M.K, Hauggaard-Nielsen H, Hogh-Jensen H, Jensen ES (2007). Competition for and utilisation of sulfur in sole and intercrops of pea and barley. Nutr Cycl Agroecosys 77: 143-153.

Andersen, M.K. (2005). Competition and complementarity in annual intercrops- the role of plant available nutrients. Department of Soil Science. The Royal Veterinary and Agricultural University, Copenhagen, Denmak.

Andow, D.A. (1991). Vegetational diversity and arthropod population response. Annual Review of Entomology 36, 561-586.

Andrews, D.J. and A.H. Kassam, (1976). The Importance of Multiple Cropping in Increasing World Food Supplies. In: Multiple Cropping, Papendick, R.I., A. Sanchez and G.B. Triplett (Eds.). American Society Agronomy, Madison, WI., USA., pp: 1-10.

Anil L, Park J, Phipps R.H., Miller F.A. (1998). Temperate intercropping of cereals for forage: A review of the potential for growth and utilization with particular reference to the UK. Grass Forage Sci 53: 301-317.

Arun Katyayan (2017). *Fundamentals of Agriculture*, Kushal Publications and Distributors Varanasi, vol. I, pp: 11-22.

Ayeni, A.O., Akobundu, I.O., Duke, W.B. (1984). Weed interference in maize, cowpea and maize/cowpea intercrop in a subhumid tropical environment. II. Early growth and nutrient content of crops and weeds. Weed Research 24, 281-290.

Banik, P., Midya, A., Sarkar, B.K., Ghose, S.S. (2006). Wheat and chickpea intercropping systems in an additive series experiment: advantages and weed smothering. Eur. J. Agron. 24, 325-332.

Bedoussac, L., Journet, E.-P., Hauggaard-Nielsen, H., Naudin, C., Corre-Hellou, G., Jensen, E.S., Prieur, L., Justes, E. (2015). Ecological principles underlying the increase of productivity achieved by cereal-grain legume intercrops in organic farming. A review. Agronomy for Sustainable Development 35, 911-935.

Beets, W.C. (1990). Raising and Sustaining Productivity of Smallholder Systems in the Tropics: A Handbook of Sustainable Agricultural Development. Agbe Publishing, Alkmaar, Netherlands.

Bodade, V.N. (1964). Mixed cropping of groundnut and Jowar. Indian oil seed J. 8(4): 297-30.

Brooker, R.W., Bennett, A.E., Cong, W.F., Daniell, T.J., George, T.S., Hallett, P.D., Hawes, C., Iannetta, P.P.M., Jones, H.G., Karley, A.J., Li, L., McKenzie, B.M., Pakeman, R.J., Paterson, E., Schoeb, C., Shen, J., Squire, G., Watson, C.A., Zhang, C., Zhang, F., Zhang, J., White, P.J. (2015). Improving intercropping: a synthesis of research in agronomy, plant physiology and ecology. New Phytologist; 206, 107-117.

Cardinale, B.J., Wright, J.P., Cadotte, M.W., Carroll, I.T., Hector, A., Srivastava, D.S., Loreau, M., Weis, J.J. (2007). Impacts of plant diversity on biomass production increase through time because of species complementarity. Proceedings of the National Academy of Sciences 104, 18123-18128.

Carrubba A, la Torre R, Saiano F, Aiello P (2008). Sustainable production of fennel and dill by intercropping. Agron Sustain Dev 28: 247-256.

Carruthers K, Prithiviraj B, Fe Q, Cloutier D, Martin RC, Smith DL (2000) Intercropping corn with soybean, lupin and forages: yield component responses. Eur J Agron 12: 103-115.

Cenpukdee U, Fukai S (1992a) Cassava/legume intercropping with contrasting cassava cultivars. 1. Competition between component crops under three intercropping conditions. Field Crops Res. 29: 113-133.

Cenpukdee U, Fukai S (1992b) Cassava/legume intercropping with contrasting cassava cultivars. 2. Selection criteria for crops. Field Crops Res 29: 135-149.

Clawson D.L. (1985). Harvest security and intraspecific diversity in traditional tropical agriculture. Econ Bot 39: 56-67.

Cong, W.F., Hoffland, E., Li, L., Six, J., Sun, J.H., Bao, X.G., Zhang, F.S., van der Werf, W. (2015). Intercropping enhances soil carbon and nitrogen. Global Change Biology 21, 1715-1726.

Dhima KV, Lithourgidis AS, Vasilakoglou IB, Dordas CA (2007) Competition indices of common vetch and cereal intercrops in two seeding ratio. Field Crops Res 100: 249-256.

El-Swaify S.A., Lo AKF, Joy R, Shinshiro L., Yost R.S. (1988). Achieving conservation effectiveness in the tropics using legume-intercrops. Soil Technol 1: 1-12.

Enyi, B.A.C. (1973). Effects of intercropping maize or sorghum with cowpeas, pigeon peas or beans Exp. Agr. 9. 83-90.

Enyi, V.A.C. (1977). Grain yield in groundnut. Exp. Agric., 13: 101-110.

Fininsa, C. (1996). Effect of bean and maize intercropping on bean common bacterial blight and rust diseases. Int. J. Pest Manage., 42: 51-54.

Francis CA, Flor CA, Temple SR (1976) Adapting varieties for intercropping systems in the tropics. In: Papendick RI, Sanchez PA, Triplett GB (eds) Multiple Cropping. ASA Special Publication 27, American Science of Agronomy, Madison, WI, USA.

Fukai S, Trenbath B.R. (1993). Processes determining intercrop productivity and yields of component crops. Field Crops Res 34: 247-271.

Grimes, A., A.M. Quasem, M.S. Uddin, N. Jahiruddin and R.N. Mallik, (1983). Performance of different cropping patterns in 1992-93 at the cropping system research site. Hathazari, Chittagong, RARS.

Hauggaard-Nielsen H, Ambus P, Jensen ES (2001). Temporal and spatial distribution of roots and competition for nitrogen in pea-barley intercrops - a field study employing 32P technique. Plant Soil 236: 63-74.

Hauggaard-Nielsen H., Jornsgaard B., Kinane J., Jensen E.S, (2008). Grain Legume-Cereal Intercropping: The practical application of diversity, competition and facilitation in arable and organic cropping systems. *Renew Agr. Food Syst.* 23: 3-12.

Horwith B. (1985). A role for intercropping in modern agriculture. BioScience 35: 286-291.

Jannasch R.W, Martin R.C (1999). The potential for capturing the forage yield of white lupin by intercropping with cereals. Biol Agr Hort 17: 113-130.

Jensen, E.S., Bedoussac, L., Carlsson, G., Journet, E.-P., Justes, E., Hauggaard-Nielsen, H. (2015). Enhancing yields in organic crop production by eco-functional intensification. Sustainable Agriculture Research 4, 42-50.

Keating B.A., Carberry P.S. (1993). Resource capture and use in intercropping - solar-radiation. Field Crops Res 34: 273-301.

Kurata, T. (1986). A study on the farming system in USSA. Q. J. Agro. Eco., 26: 179-205.

Launay M, Brisson N, Satger S, Hauggaard-Nielsen H, Corre-Hellou G, Kasynova E, Ruske R, Jensen ES, Gooding MJ (2009) Exploring options for managing strategies for pea-barley intercropping using a modeling approach. Eur J Agron 31: 85-98.

Li L, Yang S, Li X, Zhang F, Christie P (1999). Interspecific complementary and competitive interactions between intercropped maize and faba bean. Plant Soil 212: 105-114.

Li, L., Zhang, L.Z., Zhang, F.S. (2013). Crop Mixtures and the Mechanisms of Overyielding. Encyclopedia of Biodiversity, second edition. Waltham, MA: Academic Press, pp. 382-395.

Liebman, M., Dyck, E. (1993). Crop Rotation and Intercropping Strategies for Weed Management. Ecological Applications 3, 92-122.

Lithourgidis, A., Vlachostergios, D., Dordas, C., Damalas, C. (2011). Dry matter yield, nitrogen content, and competition in pea–cereal intercropping systems. Eur. J. Agron. 34, 287-294.

Lithourgidis, A.S., Dordas, C.A., Damalas, C.A., Vlachostergios, D.N. (2011). Annual intercrops: an alternative pathway for sustainable agriculture. Australian Journal of Crop Science 5, 396-410.

Litsinger, J. A. and Moody, K. (1976). Integrated pest management in multiple cropping systems. Pp. 293-317 in Papendick, R.I., Sanchez, P.A. and Triplett, G.B. (Eds) Multiple cropping. Madison, Wisconsin, American Society and Agronomy.

Lynam, J.K., J.H. Sanders and S.C. Mason, (1986). Economics and Risk in Multiple Cropping. In: Multiple Cropping Systems, Francis, C.A. (Ed.). Macmillan, New York, pp: 250-266.

Machado, S. (2009). Does intercropping have a role in modern agriculture. Journal of soil and water conservation 64, 55A-57A.

Midmore, D. J. (1993). Agronomic Modification of Resource use and Intercrop Productivity. Field Crops Research, 34: 357-380.

Mucheru-Muna M, Pypers P, Mugendi D, Kung'u J, Mugwe J, Merckx R, Vanlauwe B (2010). A staggered maize legume intercrop arrangement robustly increases crop yields and economic returns in the highlands of Central Kenya. Field Crops Res 115: 132-139.

Mugabe, N.R., M.E. Sinje and K.P. Sibuga, (1982). A Study of Crop Weed Competition in Intercropping. In: Intercropping in Semi Arid Areas, Keswani, C.L. and B.J. Ndunguru (Eds.). National Scientific Research Council and International Development Research Center, Tanzania, pp: 96-101.

Muoneke C.O, Ogwuche M.A.O., Kalu B.A. (2007). Effect of maize planting density on the performance of maize/soybean intercropping system in a guinea savannah agroecosystem. Afr J Agr Res 2: 667-677.

Nadkarni, M. V., and Vedini, K. H. (1996). Accelerating commercialization of agriculture. Economic and Political Weekly, 22, a63-a73.

Ofori, F. and W.R. Stern, (1987). Cereal legume Intercropping systems. Adv. Agron., 41: 41-90.

Ofosu-Anim J., Limbani N.V. (2007). Effect of intercropping on the growth and yield of cucumber (*Cucumis sativus* L.) and okra (*Abelmoschus esculentus* L. Moench). Int J Agr Biol 9: 594-597.

Olowe VIO, Adeyemo A.Y (2009). Enhanced crop productivity and compatibility through intercropping of sesame and sunflower varieties. Ann Appl Biol 155: 285- 291.

Palaniappan S.P, Sivaraman K. (1996). Cropping Systems in the Tropics. New Age International, New Delhi.

Plucknett D.L., Smith N.J.H. (1986). Historical perspectives on multiple cropping. In: Francis CA (ed) Multiple Cropping Systems. Mac Millan Publishing Company -New York, USA.

Ranade, C.G (1980). Impact of Cropping Pattern on Agricultural Production, *Indian Journal of Agricultural Economics*, Vol. 35, No. 2, April-June.

Reddy, M.S., Willey, R.W. (1981). Growth and resource use studies in an intercrop of pearl millet/groundnut. Field Crops Research 4, 13-24.

Santalla M, Rodino A.P., Casquero P.A, de Ron A.M. (2001). Interactions of bush bean intercropped with field and sweet maize. Eur J. Agron 15: 185-196.

Sebillotte M. (1990). In: Scaief A. (Ed.). Proceedings First Congress of European Society of Agronomy, Vol.1, Colmar, 1-16 pp.

Singh K.A., Ahlawat IPS and Mahendra Pal H. (1986). In: Proceedings National Symposium on Cropping Systems. Indian Society of Agronomy, IARI, New Delhi.

Singh, B.B. and Adjeigbe, H.A. (2002). Improving Cowpea Cereals-Based Cropping Systems in the Dry Savannas of West Africa. In: Challenges and Opportunities for Enhancing Sustain-Able Cowpea Production, Fatokun, A., S.A.Tarawali, B.B. Singh, P.M. Kormawa and M.Tamo (Eds.). IITA, Ibadan, Nigeria, pp: 276 -284.

Smith, M.E. and Francis, C.A. (1986). Breeding for multiple cropping systems. In: C.A. Francis (Editor), Multiple Cropping Systems. Macmillan, New York, pp. 219-249.

Szumigalski AR, Van Acker RC (2006). Nitrogen yield and land use efficiency in annual sole crops and intercrops. Agron J 98: 1030-1040.

Tilman, D., Balzer, C., Hill, J., Befort, B.L. (2011). Global food demand and the sustainable intensification of agriculture. Proceedings of the National Academy of Sciences 108, 20260-20264.

Trenbath, B.R. (1993). Intercropping for the management of pests and diseases. Field Crops Research 34, 381-405.

Vandermeer, J.H. (1992). The ecology of intercropping. Cambridge University Press, New York, USA.

Vyas, V. S. (1996). Diversification in Agriculture: Concept, Rationale and Approaches. *Indian Journal of Agricultural Economics*, 51(4), 636.

Wien H.C, Smithson J.B. (1981). The evaluation of genotypes for intercropping. Proceedings of the International Workshop on Intercropping, International Research Institute for the Semi-Arid Tropics (ICRISAT), 10-13 January 1979, Andhra Pradesh, India, pp. 105-116.

Willey, R.W. and M.S. Reddy, (1981). A field technique for seperating above and below ground interaction for intercropping of expt with pearl millet/groundnut. Exp. Agric., 17: 257-264.|

Willey, R.W. (1979). Intercropping: Its importance and research needs. Part 1. Competition and yield advantages. *Field Crops Abstr.* 32: 1-10.

Willey, R.W. (1990). Resource use in intercropping systems. Agric. Water Manage., 17: 215-231.

Yadav R.S, Yadav O.P. (2001). The performance of cultivars of pearl millet and clusterbean under sole cropping and intercropping systems in arid zone conditions in India. Exp Agr 37: 231-240.

Zhang F, Li L. (2003). Using competitive and facilitative interactions in intercropping systems enhances crop productivity and nutrient-use efficiency. Plant Soil, 248: 305-312.

Zhang, L.Z., van der Werf, W., Bastiaans, L., Zhang, S.P., Li, B.G., Spiertz, J.H.J. (2008). Light interception and utilization in relay intercrops of wheat and cotton. Field Crops Research 107, 29-42.

Zuofa, K., N.M. Tariah and N.O. Isinimai, (1992). Effects of groundnuts, cowpea and melon on weed control and yields of intercropped cassava and maize. Field Crops Res., 28: 309-314.

Chapter 2

A Review on *Lasiaspinosa*: Ethnobotany, Phytochemistry, Pharmacology and Karyomorphology

Dimbeshwar Das[1], Bijoy Neog[2], Ajanta Baruah Das[3] and Sandeepa Agarwalla[1]*

[1]Department of Botany, [3]IBT HUB, Gargaon College, Simaluguri – 785 686, Assam
[2]Department of Life Sciences, Dibrugarh University, Dibrugarh – 786 004, Assam
**Corresponding author e-mail: dimbeshwar@rediffmail.com*

Medicinal plants serve as a great source for the synthesis of conventional drugs. Proper analysis, awareness and knowledge about the pharmacological activities of plants is required which can be done by various experiments. From this one can ensure safety and can claim various therapeutic benefits. Lasiaspinosa is an important aroid. It is a potent plant which still needs investigation on different pharmacological activities as it contains various phytochemicals like alkaloids, phenols, tannins, saponins etc. Besides it is also a good source of dietary.

***Keywords**: Conventional, Pharmacological, Lasiaspinosa, Phytochemicals, Dietary.*

Introduction

*L. spinosa*is one of the two species of *Lasia* belonging to Araceae family (commonly known as aoids). They are locally known as Chengmora in Assam. They are also known by various other vernacular names all over India. The plant is an important group of herbal medicine since a long time as they have got many pharmacological values like anti-inflammatory, antibacterial, antihyperglycemic activity, antioxidant, *etc.* along with pharmacological values it also has a wide range of ethnobotanical significance. They are also considered as an indigenous plant because of their high nutritional contents but it is still an underutilized vegetable.

Lasiaspinosa is a large marsh plant that is aquatic or terrestrial, shot-stemmed with underground rhizome. They mostly occur in wet forests, open marshes, wetlands or in permanently standing water. It is mostly distributed in South Asia and South-East Asia (Keating and Gregory, 2002).

Ethnobotany

Most of the aroids are used as traditional food and in medicine. Young tender leaves of *L. spinosa* are used as vegetable (Johnson, 1999). The peeled leaf stalks are edible and are used as flavoring agent in curry. The rhizoids are used to cure snake bites, gastritis and inflammation. The leaves are used for treating coughs, stomachaches and joint pains. In the folk medicine of the Naga tribes of India, the porridge made from young tender leaves of young tender leaves is frequently used to treat intestinal worm infections. The pudding is locally known as "jurang" (Temjenmongla and Yadav, 2006). According to the bodo community these plant is believed to be the cheapest source of multivitamins (Brahma *et al.*, 2014). Tripuri medicine practitioner uses the paste made from the leaves of the *L. spinosa* in applying to wounds and burn skin (Sen *et al.*, 2011). Leaves and corms are given as a medication for piles in Khagrachari (Yusuf *et al.*, 2009). In Rajshahi and Natore district of Bangladesh. The tuber of *Lasiaspinosa* (L.) Thwaites are recommended for treatment of rheumatoid arthritis, constipation, and also to purify bloods (Hossain *et al.*, 2010). Besides being useful to human, active phytochemicals present in *L. spinosa* would help to obtain a natural growth promoter for sustainable aquaculture farming (Munglue *et al.*, 2019). The active compounds present in L.*spinosa* play an important role in ant stress, growth stimulation, appetite activation and immunomodulation in aquatic animals (Citarasu, 2010; Leal *et al.*, 2016; Couto *et al.*, 2014).

Phytochemistry

From the various reports published, the main chemical present in *L. spinosa* includes alkaloids, saponins, phenols, tannins, flavonoids and terpenoids. Besides the mentioned secondary metabolites, they also contain good amount of carbohydrates, proteins, fats, lignin and glycosides that were experimented by the phytochemical screening of leaf extract of *L. spinosa* done in different solvents (Kumar *et al.*, 2013). Stems of *Lasiaspinosa* also showed presence of pro-vitamin A and carotenoid (Priyadarshani, 2007). Sterols are also present in *L. spinosa* (Firdusi *et al.*, 2012).

The terpenes predominate in the plants of Araceae family. Aerial parts of *Lasiaspinosa* (L.) Thwaites may have a good correlation to the bioactivities of the plants. The results of some other studies showed the presence of six compounds namely α-pinene (68.71 per cent), camphene (3.76 per cent), δ-3-carene (15.19 per cent), caryophyllene (4.98 per cent) limonene (2.13 per cent) and α-selinene (3.23 per cent) and all these six compounds comprised of 98.0 per cent of essential oil (*EO*) resulting in bioactivity as well as leading to pharmacological benefits to human health (Magwa *et al.*, 2006; Leguet *et al.*, 2014).

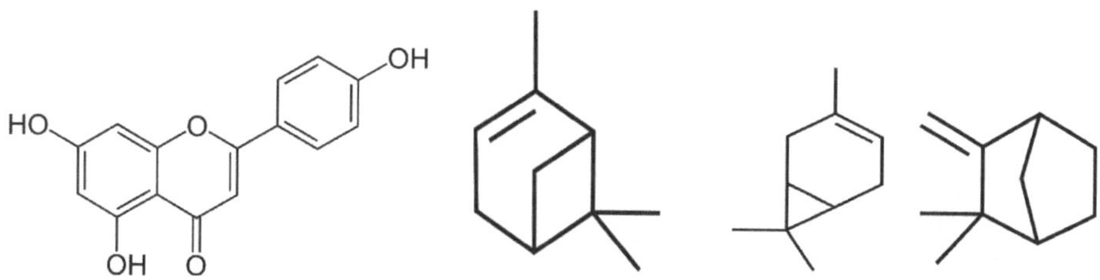

Apigenin α- pinene **Camphene** **3-carene**

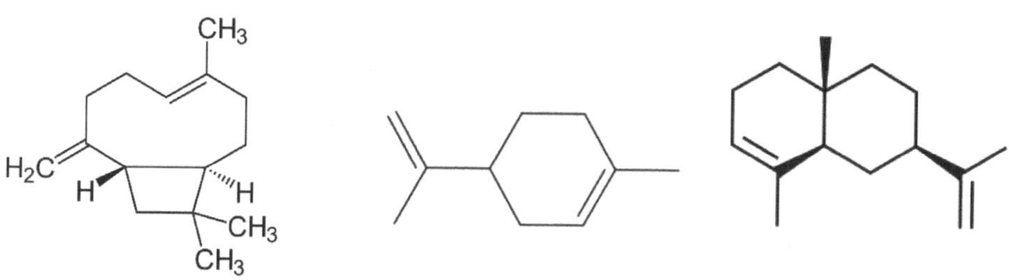

Caryollene **Limonene** **α-selinene**

Figure 2.1: Structures of Terpenoids.

One of the investigation showed presence of phenolic contents and the volatile compounds were methyl ester of oleic acid, palmitic acid and stearic acid as determined by GC/MS. Phenolic contents of the study were determined qualitatively and quantitatively by HPLC/TOF-MS. Six phenolic compounds (syringic acid, morin, gentistic acid, 4-hydroxybenzoic acid, cinnamic acid, and apigenin) were found in the extracts. GC/MS analysis of steam distilled essential oil showed camphor, α-pinene and δ-3-carene as the main constituents (Rahman *et al.*, 2019).

Another study indicated that *L. spinosa* consists of 17.6 kcal/100g protein, 83 kcal/100g moisture, 1.16 kcal/100g fats, 34 kcal/100g ash, 17 kcal/100g total solids, 35.7 kcal/100g carbohydrate. Micronutrients such as Zinc, Magnesium, Molybdenum, Copper, Iron and Manganese are also present in 7.44 ppm, 6.22 ppm, 1.18 ppm, 0.31 ppm, 17.06 ppm, 1.33 ppm respectively[4]. Rhizome is also a rich source of dietary fiber with 40 per cent -75 per cent of total dietary fiber on dry weight basis (Shefana and Ekanayake, 2009).

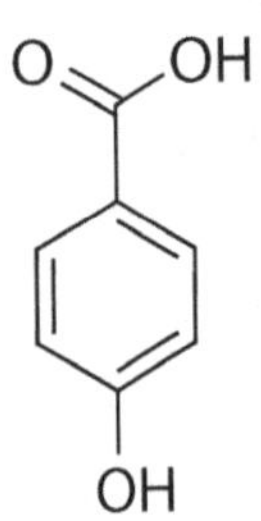

4-hydroxybenzoic acid

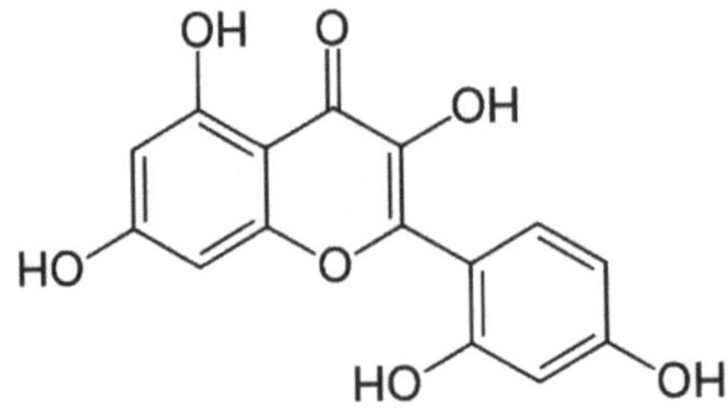

Morin

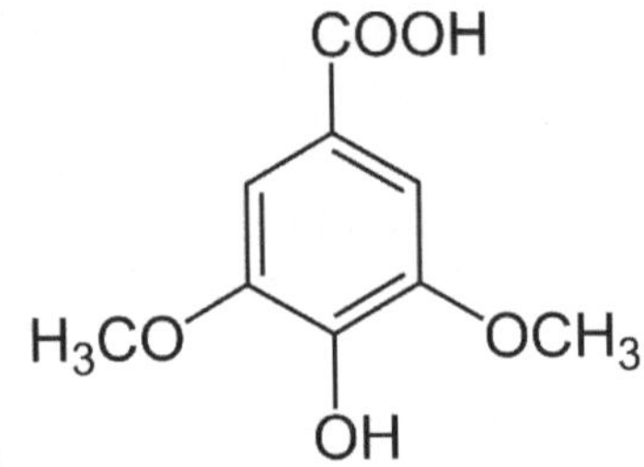

Syringic acid

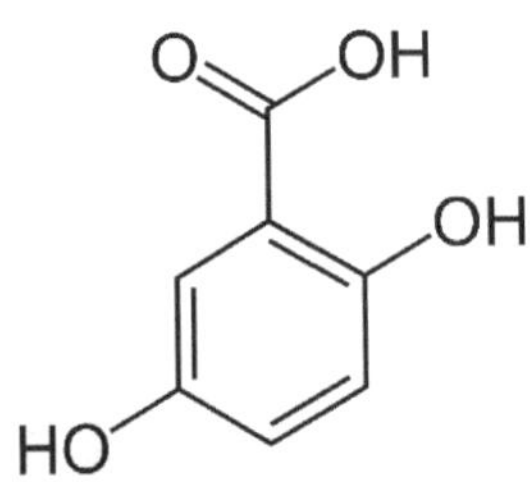

Gentistic acid

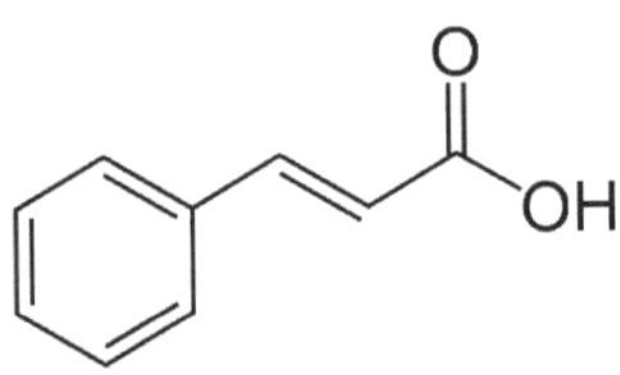

Cinnamic acid

Figure 2.2: Structure of Phenolic Compounds Present in *Lasia spinosa*.

Pharmacological Effects

Plant secondary metabolites are very important for the plant as they provide mechanical defense to the plant and protect them from predators, besides, the chemicals are very important from the point of view of pharmacological industry because of the various bioactivity that benefits human.

In one of the experiments anthelmintic activity of *L. spinosa* was studied from three different concentrations (25 mg/ml, 50 mg/ml and 100 mg/ml) of methanol extract. The activity determined the time taken for paralysis and death of worms. The experiment indicated that methanol extract significantly exhibited paralysis and also caused death of worms especially at highest concentration of 100 mg/ml. The study concluded that it has got potent anthelmintic effect (Goshwami *et al.*, 2013). Trichinellosis is a widespread disease caused by *Trichinella spiralis* for which no effective drug treatment is available. One of the studies showed that leaf extracts of *L. spinosa* is effective against trichinellosis more than the drugs albendazole and malbendazole (Yadav, 2012).

Another study showed that *L. spinosa* has antihyperglycemic effect. To determine the antihyperglycemic effects, methanolic and ethanolic extracts of leaf of *L. spinosa* plant were experimented in Swiss albino mice. Different dosses (200 and 400mg/kg bw-weight) for each extracts were orally administered. Significant hypoglycemic activity was observed regarding both of the extracts and dosses compared to control when serum was measured by glucometer. The methanolic extract showed the higher inhibitory activity than ethanolic extract. Thus, the result pointed out that the extracts of *L. spinosa* are antihyperglycemic (Hasan *et al.,* 2014).

The results of a recent study indicated that ethanol extracts of *L. spinosa* leaves showed anti-oxidant activity. The antioxidant activity is observed due to the presence of high levels of polyphenolic compounds. Treatment with the leaf extract significantly repressed the production of inflammatory mediators that confirmed its anti-inflammatory effect (Nguyen *et al.,* 2020). In different study anti-nociceptive, anti-inflammatory, and anti-diarrheal activities of *L.spinosa* in rodents from hydro alcoholic extract of roots was performed that supported its activity (Deb *et al.,* 2010).

From an investigation it was concluded that the various partitionates of methanolic extracts of *L. spinosa* leaves showed significant antinociceptive effects. The investigation was performed using both acetic acid writhing method and radiant heat tail flicking method. Petroleum ether fraction showed significant effect while the least was shown bydichloromethane (Goshwami *et al.,* 2012). Another observation concluded that L.*spinosa* leaf extract possesses an anticestodal property which is validated and supported by the reduction in EPG counts and worm recovery rates in extract-treated rats (Temjenmongla and Yadav, 2006).

From another analysis of an experiment it was concluded that the compounds present in *L. spinosa* are compatible as a strong antioxidant against pathogens in different doses. This was found by performing computational screening of the rhizomes of *L. spinosa* (Marisetti *et al.,* 2020). The leaves extract of *L. spinosa* showed mild antimicrobial property (Goshwami *et al.,* 2012).

One of the studies conducted in Sri Lanka proved that polyphenols and ascorbic acid contributes majorly to the total antioxidant activity of the rhizome of *L. spinosa* (Shefana and Ekanayake, 2009). The biochemical parameters such as Total Cholesterol, Triglycerides, LDL-C, HDL-C, VLDL-C, Serum Amylase, Serum Lipase and WBC Count were estimated to evaluate the anti-hyperlipidemic activity from the different parts of the *L. spinosa*, from which it was found that the leaves have anti-hyperlipidemic activity. The presence of alkaloids, glycosides, saponins, tannins, proteins and flavonoids and ascorbic acid were screened from phytochemical tests. The results were supported by histopathological evaluation of liver and pancreas (Mahmood *et al.,* 2015).

Bioactive fraction from *L. spinosa* rhizome on experimentally induced gastric ulceration revealed that the plant has gastro-protective potentials (Dubey *et al.,* 2014; Neamsuvan and Ruangrit, 2017). The presence of phenolic compound and high percentage of gallic acid proves the potentiality. Gallic acid because of its activities like scavenging free radicals, blocking •OH-mediated oxidative damage, plays important role in the prevention and therapy of diseases especially as an antioxidant (Kankanamge *et al.,* 2017).

Karyomorphology

In an investigation of karyomorphological variations in three morphologically different forms showed that the sagittate form had 2n=27 numbers of chromosome with a minute subtelocentric chromosome. The centromeric formula of sagittate form was 19m+7sm+1st. While both lamina dissected and mixed form possessed 2n=26 chromosomes with centromeric formula 9m+15sm+2st in the lamina dissected form and 14m+11sm+1t in the mixed form. The report also showed the total length of 2n chromosome complement as 102.23 mm in the sagittate form, 57.78 mm in the lamina dissected form and 74.46 mm in the mixed form. They also found one satellite in sagittate form, 2 in the lamina dissected form and 5 in the mixed form in orcein staining and in CMA-staining, 2 big CMA-positive satellites were found in the sagittate form. Two small CMA-positive satellites were found in the lamina dissected form while the mixed form had 1 big and 1 small satellite. Two CMA-positive bands were found in sagittate, 5 in lamina dissected and 3 in mixed form. The percentages of GC-rich region were respectively 2.88, 11.23 and 6.79 in the sagittate, lamina dissected and the mixed form respectively. The study also postulated that the mixed form might be a natural hybrid between the sagittate and the lamina dissected form (Sultana *et al.,* 2006).

A karyomorphological study on two morphological forms of *L. spinosa* collected from Assam revealed that there is a difference in chromosome number between the two forms. In the mixed form the number of chromosome was found to be 2n=26 in somatic cells with chromosome length varying from 2.31um-4.08um while their volumes ranged from 0.034um-3.70um.On the other hand in the lamina dissected form number of chromosomes was found to be 2n=24 in somatic cells with chromosome length ranging from 1.09um-4.42um and their volumes varying from 0.015um3-0.40um^3. The report also believed that their variation from a previous report might be due to ecological variations (Hore and Tanti, 2014).

In another cytogenetical investigation based on the colour of spathe, namely the common red flower *L. spinosa* and its rare Green flower form,numbers of chromosomes were found to be 2n=26 and 2n=28 respectively. RAPD analysis was also performed to elucidate the taxonomic rank of the green flower form. The investigation revealed two sub-metacentric chromosomes in the green flower form. In contrast no sub-metacentric chromosome was found in the red flower form. Only in the red flower form two stain specific satellites were found. No CMA-band was found in the red flower form. On the other hand, 3 prominent CMA-positive bands were found in the green flower form. The 2 forms showed distinct RAPD markers with 5 primer combinations concluding that the two forms are entirely morphologically different (Alam *et al.,* 2012).

REFERENCES

Alam, S. S., Munira, S., Habib, M. A., and Sultana, S. S. (2012). Karyotype and RAPD analysis in common and rare forms of Lasia spinosa (L.) Thwaites. *Cytologia, 77*(4), 499-505.

Brahma, J., Chakravarty, S., and Rethy, P. (2014). Qualitative estimation of the presence of bioactive and nutritional compound in *Lasiaspinosa*: an important vegetable plant used by the Bodos of Kokrajhar District. *Int. J. Chem. Tech. Res, 6*, 1405-1412.

Citarasu, T. (2010). Herbal biomedicines: a new opportunity for aquaculture industry. *Aquaculture International, 18*(3), 403-414.

Couto, A., Kortner, T. M., Penn, M., Bakke, A. M., Krogdahl, Å., and Oliva-Teles, A. (2014). Effects of dietary phytosterols and soy saponins on growth, feed utilization efficiency and intestinal integrity of gilthead sea bream (Sparus aurata) juveniles. *Aquaculture, 432*, 295-303.

Deb, D., Dev, S., Das, A. K., Khanam, D., Banu, H., Shahriar, M., and Basher, S. A. M. K. (2010). Antinociceptive, anti-inflammatory and anti-diarrheal activities of the hydroalcoholic extract of *Lasiaspinosa* Linn. (Araceae) roots. *Latin American Journal of Pharmacy, 29*.

Dubey, M. K., Das, S., Yadav, S., Gupta, P. C., Jaiswal, S. K., and Sharma, B. (2014). Gastroprotective potential of bioactive fraction from *Lasiaspinosa* rhizome on experimentally induced gastric ulceration. *International Journal of Pharmaceutical Sciences and Research, 5*(12), 5209.

Firdusi, S. N., Neog, N., and Barooah, M. S. (2012). Phytochemical analysis of *Lasiaspinosa* and Alpinianigra, potential medicinal plants of Assam. *Medicinal Plants-International Journal of Phytomedicines and Related Industries, 4*(3), 170-173.

Goshwami, D. U. R. A. J. A. N., Rahman, M. M., Muhit, M. A., and Islam, S. (2013). *In vitro* evaluation of anthelmintic activity of *Lasiaspinosa* leaves. *Int J Curr Pharm Res, 5*(1), 34-35.

Goshwami, D., Rahman, M. M., Muhit, M. A., and Islam, M. S. (2012). Antinociceptive activity of leaves of *Lasiaspinosa*. *Archives of Applied Science Research, 4*(6), 2431-4.

Goshwami, D., Rahman, M. M., Muhit, M. A., Islam, M. S., and Anasri, M. (2012). Antioxidant property, cytotoxicity and antimicrobial activity of *Lasiaspinosa* leaves. *Nepal Journal of Science and Technology, 13*(2), 215-218

Hasan, M. N., Munshi, M., Rahman, M. H., Alam, S. N., and Hirashima, A. (2014). Evaluation of antihyperglycemic activity of *Lasiaspinosa* leaf extracts in Swiss albino mice. *World J. Pharm. Pharm. Sci, 3*, 118-124.

Hore, P., and Tanti, B. (2014). Karyomorphological studies of two morphotypes of Lasiaspinosa (Lour.) Thwaites available in Assam, India. *Annals of Plant Sciences, 3*(08), 792-796.

Hossain, S., Khatun, A., and Miajee, U. E. (2010). Medicinal plants used by folk medicinal practitioners in three villages of Natore and Rajshahi districts, Bangladesh. *American-Eurasian Journal of Sustainable Agriculture, 4*(2), 211-218.

Johnson T. CRC ethnobotany desk reference, CRC press, Bocn Raton, 1999, pp.458

Kankanamge, S. U., Amarathunga, A. A. M. D. D. N., Sanjeewani, N. A., and Samanmali, B. L. C. (2017). Phytochemical and ethno-pharmacological properties of Lasiaspinosa (Kohila): A review. *World J. Pharm. Res, 6*, 1-9.

Keating R.C and Gregory M. Acoraceae and Araceae.USA: Oxford University Press.2002

Kumar, M., Mondal, P., Borah, S., and Mahato, K. (2013). Physico-chemical evaluation, preliminary phytochemical investigation, fluorescence and TLC analysis of leaves of the plant *Lasiaspinosa* (Lour) Thwaites. *Int J Pharm Pharm Sci, 5*(2), 306-310.

Leal, M. C., Ferrier Pagès, C., Petersen, D., and Osinga, R. (2016). Coral aquaculture: applying scientific knowledge to *ex situ* production. *Reviews in Aquaculture, 8*(2), 136-153.

Leguet, A., Gibernau, M., Shintu, L., Caldarelli, S., Moja, S., Baudino, S., and Caissard, J. C. (2014). Evidence for early intracellular accumulation of volatile compounds during spadix development in Arum italicum L. and preliminary data on some tropical Aroids. *Naturwissenschaften, 101*(8), 623-635.

Magwa, M. L., Gundidza, M., Gweru, N., and Humphrey, G. (2006). Chemical composition and biological activities of essential oil from the leaves of Sesuvium portulacastrum. *Journal of Ethnopharmacology, 103*(1), 85-89.

Mahmood, S. B., Atif, M., Ali, S. R., Ahmed, M. I., and Rahman, S. A. (2015). Evaluation of Antihyperlipidemic Activity of Methanolic Leaves Extract of *Lasiaspinosa* and Its Role in Prevention of Hyperlipidemia Induced Pancreatitis in Rats. *International Journal of Pharmaceutical Sciences and Research, 6*(4), 1502.

Marisetti, A. L., Bonthu, M. G., and Battu, G. R. (2020). Antioxidant and Anti-inflammatory Activity Screening of *Lasiaspinosa* Rhizome and its Validation using a Computational Simulation Approach. *Indian Journal of Pharmaceutical Education and Research, 54*(4), 1109-1120.

Munglue, P., Rattana, K., Sangchanjiradet, S., and Dasri, K. (2019). Effect of dietary *Lasia* (*Lasiaspinosa* (L.) Thwaites) extract on growth performance and intestinal histology in hybrid catfish (Clarias macrocephalus× Clarias gariepinus). *Chiang Mai University Journal of Natural Sciences, 18*(2), 226-249.

Neamsuvan, O., and Ruangrit, T. (2017). A survey of herbal weeds that are used to treat gastrointestinal disorders from southern Thailand: Krabi and Songkhla provinces. *Journal of ethnopharmacology, 209*, 318-327.

Nguyen, T. Q., Duy Binh, T., Pham, T. L., Nguyen, Y. D., Dai Trang, T. X., Nguyen, T. T., and Kamei, K. (2020). Anti-inflammatory effects of *Lasiaspinosa* leaf extract in lipopolysaccharide-induced RAW 264.7 macrophages. *International journal of molecular sciences, 21*(10), 3439.

Priyadarshani, A. M. B. (2007). *Studies on carotenoid content of selected fruits and non-leafy vegetables of Sri Lanka and their bioavailability and metabolites* (Doctoral dissertation, University of Sri Jayewardenepura, Nugegoda).

Rahman, A., Siddiqui, S. A., Oke-Altuntas, F., Okay, S., GÜL, F., and Demirtas, I. (2019). Phenolic profile, essential oil composition and bioactivity of *Lasiaspinosa* (L.) thwaites. *Brazilian Archives of Biology and Technology, 62*.

Sen, S., Chakraborty, R., De, B., and Devanna, N. (2011). An ethnobotanical survey of medicinal plants used by ethnic people in West and South district of Tripura, India. *Journal of Forestry Research, 22*(3), 417-426.

Shefana, A. G., and Ekanayake, S. (2009). Some nutritional aspects of *Lasiaspinosa* (kohila).

Sultana, S., Rahman, R. B., and Alam, S. S. (2006). Karyotype analysis in three morphological forms of Lasia spinosa (L.) Thwaites (Araceae). *Cytologia, 71*(4), 359-364.

Temjenmongla, T., and Yadav, A. K. (2006). Anticestodalefficacy of *Lasiaspinosa* extract against experimental Hymenolepis diminuta infections in rats. *Pharmaceutical Biology, 44*(7), 499-502.

Yadav, A. K. (2012). Efficacy of *Lasiaspinosa* leaf extract in treating mice infected with Trichinella spiralis. *Parasitology research, 110*(1), 493-498.

Yusuf, M., Begum, J., Hoque, M. N., and Chowdhury, J. U. (2009). Medicinal Plants of Bangladesh, Bangladesh Council of Scientific and Industrial Research. *Dhaka, Bangladesh*.

Chapter 3

Frequency of Filamentous Fungi on Woody Litter of Mangroves along North Malabar, Kerala, India

R. Nambiar Gayatri and K. Shilpa*

Department of Post Graduate Studies and Research in Botany, Sir Syed College, Taliparamba, Kannur – 670 142, Kerala

**Corresponding author e-mail: grn_149@rediffmail.com*

Decomposition of organic material, mainly wood, by marine fungi has recently become a focal point of research. With the advent of ecology of these organisms a better understanding of their role and function is now emerging. During a marine mycological survey on mangrove substrates along Kerala coast, 48 manglicolous marine fungi comprising 33 ascomycetes, 2 basidiomycetes and 13mitosporic fungi were encountered on various mangrove hosts viz. Acanthus ilicifolius, Acrostichium, Avicennia, Bruguiera, Kandeliacandel and Rhizophora. Fungal composition varied from one host to another. Each host had its own most frequent, frequent, occasional and rare species.

Keywords: *Mangroves, Marine fungi, Per cent frequency of occurrence.*

Introduction

Mangrove forests generate considerable amount of detritus such as leaf litter, woody debris and inflorescence. They constitute an ideal environment for many detritus dependent microorganism. As fungal population is involved in detritus processing a considerable alternation is necessary for assessment of mycoflora (Maria and Sridhar, 2005). Mangrove inhabiting marine fungi are known as manglicolous marine fungi. Although there are salt sprays, salt water never reaches the upper part of mangrove trees. But the trunks and aerating roots are permanently or intermittently submerged. Owing to this, terrestrial fungi and lichens are seen on the upper part of mangrove trees, marine fungi on the lower part and an overlap of terrestrial fungi with the marine fungi in the middle (Kohlmeyer and Kohlmeyer, 1979).

Mangrove inhabiting marine fungi were first reported by Cribb and Cribb (1955). Studies reveal that fifty five mangroves and their associates yielded about 200 higher marine fungi. Among the tree members *Rhizophora* and *Avicennia* harbours maximum number of higher marine fungi (Sarma and Vittal, 2000; Raveendran and Manimohan, 2007). Diversity of mangrove fungi depend certain factors, which include age of the mangrove, diversity of mangrove plant species, physiochemical features of mangrove habitat like temperature, salinity and tidal range (Hyde and Jones, 1988). Marine fungi occur on prop roots, pneumatophores, branches and drift wood in the intertidal region of mangrove stand but are absent on substrates embedded in the mud (Kohlmeyer and Kohlmeyer, 1979).

A variety of micro and macro ecological niches in mangrove ecosystem promote to successful existence of wide range of mycoflora. Senescent mangrove leaves detach from plant and reach different distinct habitats as follows – they may remain entrapped above the water level in certain region which is not accessible for tides, hence gets dried and reach water by storms or wind, during high tide conditions they may float on the surface of water, during low tide situation they may reach the mangrove floor, may entrapped at some depth below the water surface, or gets settle on the sediment or gets trapped in the sediment. In the same manner twigs dried during summer gets separated by heavy rains of monsoon season are entangled in canopy later they will enter water with or without bark and settle on floor or remain in settlement, stems, prop roots and pneumatophores may decay but still remain attached to parent plant. These conditions favour the colonization and activity of marine fungi in mangrove habitats (Newell, 1996).

The season in which detritus are recovered from the mangroves decides fungal fungal composition. Monsoon begins in southwest coast of India in late May or early June and continues up to September. Heavy rainfall in the Western Ghats results in flushing freshwater and sediments to mangrove habitats and salinity declines to zero. During post monsoon until January, salinity raises to 50 per cent (Sridhar and Kaveriappa, 1988). This facilitates colonization of freshwater fungi on mangrove substrates during rainy season. Live and dead twigs of mangrove canopy harbor terrestrial fungi. When such substrata get into mangrove waters during monsoon,these fungi dominate at least for a few months (Maria and Sridhar, 2003). Although there are earlier reports about marine mycology in Kerala coast, it is not so exclusive with the mangrove host. Hence, the present study was undertaken.

Materials and Methods

Collection and Treatment of Wood Samples

Decaying mangrove samples (*Avicennia, Bruguiera, Rhizophora, Kandeliacandel, Acanthus ilicifolius* and *Acrostichium*) were collected from the mangrove ecosystem of North Malabar region of Kerala (Dharmadam, Telicherry, Mellor, Nadal, Madakkara, Pazhayangadi and Valapattanam) during April 2016 –September 2018. They were washed well, placed in sterile polythene bags and were brought to the laboratory. Samples were observed under stereomicroscope for fungal fructifications and incubated in polythene bags at room temperature. Periodical isolation of marine fungi from these samples was carried out for one year. Identifications of marine fungi were done using taxonomic keys by Kohlmeyer and Kohlmeyer, (1979), Kohlmeyer and Volkmann Kohlmeyer (1991), Sarma and Vittal (2004), Raveendran and Manimohan (2007) and Gayatri and Raveendran (2013). The marine fungi thus identified were tabulated and recorded (Table 3.1).

Table 3.1: Per cent Frequency of Marine Fungi on Various Hosts along North Malabar Region of Kerala

Name of Marine Fungi	*Acan*	*Acro*	*Avi*	*Bru*	*Kan*	*Rhi*
Ascomycetes						
Aigialus grandis Kohlm *et* Schatz	0.00	0.00	7.34	0.00	0.00	0.00
Aigialus mangrovei Borse	0.00	0.00	3.67	5.56	3.23	0.00
Aigialus parvus Schatz *et* Kohlm.	0.00	0.00	2.75	1.85	0.00	3.57
Aniptodera chesapeakensis Shearer *et* Mill	0.00	0.00	0.92	3.70	0.00	12.50
Aniptodera indica Ananda *et* Sridhar	5.56	0.00	2.75	1.85	0.00	0.00
Aniptodera longispora Hyde	0.00	0.00	3.67	0.00	0.00	8.93
Aniptodera mangrovei Hyde	0.00	0.00	5.50	11.11	0.00	7.14
Anthostomella nypensis Hyde, Alias *et* Lu	0.00	0.00	0.92	0.00	0.00	7.14
Astrosphaeriella striatispora (Hyde) Hyde	0.00	0.00	3.67	0.00	0.00	0.00
Carinispora velatispora Hyde	0.00	0.00	1.83	0.00	0.00	0.00
Cryptospha eriamangrovei Hyde	0.00	0.00	1.83	0.00	0.00	0.00
Dactylospora haliotrepha Kohlm *et* (Kohlm.) Hafellner	0.00	0.00	6.42	3.70	3.23	10.71
Didymosphaeria maritima (P. Crouan and H. Crouan) Sacc.	0.00	0.00	0.92	0.00	0.00	0.00
Eutypaba thurstensis Hyde *et* Rappaz	0.00	0.00	17.43	5.56	0.00	16.07
Halasarpheia kandeliae Abdel - Wahab *et* Jones	0.00	0.00	0.92	0.00	6.45	0.00

Name of Marine Fungi	Acan	Acro	Avi	Bru	Kan	Rhi
Halorosellinia oceanica (Schatz) Whalley, Jones, Hyde *et* Laessoe	0.00	0.00	1.83	1.85	12.90	3.57
Halosarpheia marina (Cribb et. Cribb) Kohlm	0.00	0.00	3.67	0.00	0.00	0.00
Halotthia posidoniae (Durieu *et* Montagne in Montagne) Kohlm	0.00	0.00	0.00	0.00	0.00	1.79
Herpotrichea nypicola Hyde *et* Alias	0.00	0.00	4.59	0.00	0.00	0.00
Hypophloeda sp.	0.00	0.00	0.92	0.00	0.00	0.00
Leptosphaeria australiensis (Cribb *et* Cribb.) Hughes.	27.78	0.00	4.59	1.85	3.23	0.00
Leptosphaeria peruviana Speg. in *An. Soc. Cient. Argent.* 12 : 179, 1881.	0.00	0.00	0.00	5.56	0.00	1.79
Lignin colalaevis Hohnk	22.22	0.00	2.75	11.11	6.45	0.00
Lineolata rhizophorae (Kohlm et Kohlm) Kohlm *et* Kohlm	0.00	0.00	0.00	12.96	0.00	28.57
Lulworthia grandispora Meyers	0.00	0.00	5.50	5.56	0.00	5.36
Lulworthia kniepii (Ade and Bauch) Petr.	0.00	0.00	0.00	3.70	0.00	0.00
Marinosphaera mangrovei Hyde	11.11	0.00	1.83	1.85	0.00	0.00
Morosphaeriara munculicola (Hyde) Suetrong, Sakayaroj, Jones *et* Schoch.	0.00	0.00	0.00	0.00	0.00	7.14
Savoryella lignicola Jones *et* Eaton	22.22	0.00	8.26	0.00	0.00	0.00
Savoryella longispora Jones *et* Hyde	0.00	0.00	7.34	0.00	9.68	0.00
Savoryella paucispora (Cribb *et* Cribb) Koch	0.00	0.00	0.00	0.00	0.00	3.57
Verruculinia enalia Kohlm) Kohlm *et* Volkm Kohlm	0.00	0.00	12.84	5.56	12.90	14.29
Zopfiella sp.	0.00	0.00	0.00	0.00	0.00	3.57
Basidiomycetes						
Calathella mangrovei Jones *et* Agerer	0.00	0.00	1.83	0.00	0.00	0.00
Halocyphina villosa Kohlm	0.00	0.00	9.17	3.70	29.03	1.79
Mitosporic Fungi						
Bactrodesmium linderi (Crane *et* Shearer) Palm *et* Stewart	0.00	6.67	0.00	0.00	0.00	8.93
Clavatospora bulbosa (Anast.) Nakagiri *et* Tubaki	27.78	0.00	2.75	0.00	9.68	0.00
Halenosporavaria (Anastasiou) Jones	33.33	20.00	0.92	0.00	0.00	0.00
Helicorhoidionnypicola Hyde and Goh.	0.00	0.00	0.00	0.00	0.00	1.79
Monodictys pelagica (Johnson) Jones	0.00	13.33	0.00	0.00	0.00	10.71

Name of Marine Fungi	*Acan*	*Acro*	*Avi*	*Bru*	*Kan*	*Rhi*
Tetra ploaaristata Berk *et* Br	0.00	0.00	0.00	1.85	0.00	7.14
Trichocladium achrasporum (Meyers *et* Moore) Dixon	0.00	0.00	1.83	0.00	0.00	0.00
Trichocladium alopallonellum (Meyers *et* Moore) Kohlm *et* V. Kohlm	0.00	13.33	6.42	14.81	0.00	12.50
Trichocladium constrictum Schmidt	0.00	0.00	0.92	0.00	19.35	3.57
Trichocladium lignicola Schmidt	0.00	0.00	0.92	3.70	0.00	0.00
Trichocladium melhae Jones, Abdel-Wahab *et* Vrijmoed	0.00	20.00	1.83	11.11	6.45	0.00
Trichocladium nypae Hyde *et* Goh	0.00	0.00	0.00	7.41	0.00	1.79
Zalerion maritimum (Linder) Anastasiou	0.00	0.00	0.00	3.70	12.90	1.79

Acan: *Acanthus ilicifolius; Acro*: *Acrostichium; Avi*: *Avicennia; Bru*: *Bruguiera, Kan*: *Kandeliacandel* and *Rhi*: *Rhizophora.*

Presentation of Data

Percent frequency of occurrence (FO) = Total number of isolates of a particular species divided by total number of samples supporting marine fungi X 100.

On the basis of percentage frequency of occurrence, the marine fungi were classified as most frequent (>10 per cent), frequent (5-10 per cent), occasional (3-5 per cent) and rare (<3 per cent).

Average number of fungi per sample = Total number of occurrences of marine fungi divided by total number of samples colonized by marine fungi

Results and Discussion

A total of 48 manglicolous marine fungi comprising 33 ascomycetes, 2 basidiomycetes and 13mitosporic fungi were encountered on various mangrove hosts *viz. Acanthus ilicifolius, Acrostichium, Avicennia, Bruguiera, Kandeliacandel* and *Rhizophora*. Fungal composition and frequency varied from one host to another. Maximum numbers of species (35) were recorded from *Avicennia* and minimum from *Acrostichium* (5). Average number of marine fungi was highest for *Rhizophora* (2.5)and least for *Acrostichium* (0.5).

Of the seven marine fungal species obtained from *Acanthus ilicifolius* 5 were ascomycetes and 2 were mitosporic fungi. Except *Aniptodera indica* all others were of the most frequently encountered group. Five mitosporic species were obtained from *Acrostichium. Bactrodesmium linderi* was found to be comparatively more dominant.

Thirty five species including 26 ascomycetes, 2 basidiomycetes and 7mitosporic fungi were identified from *Avicennia. Eutypa bathurstensis* (17.43 per cent) and *Verruculina enalia* (12.84 per cent) were most frequent in occurrence, seven species were frequent, six species were occasional and nineteen species were rarely encountered from *Avicennia*. Twenty three species were isolated from *Bruguiera* which includes 16 ascomycetes, 1 basidiomycete and 6mitosporic fungi. Five species were most frequently isolated, six species each were frequently, occasionally and rare in occurrence from *Bruguiera*.

Thirteen species were identified from *Kandeliacandel*. These include 8 ascomycetes, 1 basidiomycete and 4 mitosporic fungi. Among the five most frequent fungi *Halocyphina villosa* (29.03 per cent) was most dominant. Five species were frequent and 3species were occasionally seen in *Kandeliacandel*. Of the twenty five species obtained from *Rhizophora*, 16 were ascomycetes, 1 was basidiomycetes and 8were mitosporic fungi. Seven species each were most frequent and frequently encountered, five species were occasional and six species were rare in occurrence.

Certain fungi encountered in the present study were previously reported from restricted hosts. However, current study supports the records of many species from new hosts thereby extending the host range of these species. *e.g. Lineolatarhizophorae* was recorded for the first time from *Bruguiera; Saccardoella mangrovei* from *Kandeliacandel*; *Halosarpheia kandeliae* from *Avicennia*. A range of fungi occur in the marine ecosystem although these differ as to their location, host and some fungi occur more frequently than others. Earlier reports by previous workers (Hyde and Jones, 1988; Ravikumar and Vittal, 1987; Sarma and Vittal, 2000; Borse, *et al.*, 2000; Sarma, *et al.*, 2001; Maria and Sridhar, 2002,2005, 2006a, b; Gayatri *et al.*, 2006 and Raveendran and Manimohan 2007, Gayatri and Raveendran, 2012, 2013b, c, 2014) support the same observation. However, the frequency of occurrence and relative abundance of marine fungi from various mangrove forests of world shows variation. This could be attributed to the difference in the species diversity of mangrove ecosystem, age and preference of host substrate, ecological factors *etc.* which support the distribution of marine fungi.

Though studies conducted on marine fungi have so far been on taxonomy, biodiversity and ecology there is still an urgent need to document the mycota of coastal areas from different parts of the world. Moreover, coastal ecosystem is being destroyed by a number of anthropogenic activities. Mangroves and other coastal areas are vulnerable to the these disturbances This will in turn effect the ecological role of marine fungi *i.e.* depletion of marine microflora especially mycoflora will gradually leads to the depletion or ill- healthy growth of mangrove vegetation and finally leads to erosion and destruction of mangrove forest, fish population *etc.* All this will directly or indirectly effect the employment, food security and economy of the country. Hence, it is suggested that microbial diversity studies in addition to floral and faunal diversity studies need to be undertaken with high priority from coastal ecosystem (Gayatri and Raveendran, 2012, 2013b).

Acknowledgments

The authors are thankful to the Principal and Management of Sir Syed College, Taliparamba for providing facilities.

REFERENCES

Borse, B. D., Kelkar, D. J. and Patil, A. C. Frequency of occurrence of marine fungi from Pirotan Island (Gujarat), India. *Geobios.*2000, 27, 145 -148.

Cribb, A.B. and Cribb, W. Marine fungi from Queensland – 1.*Univ. Queensl.Pap.Dep. Bot.*, 1955, 3, 77- 81.

Gayatri, R. Nambiar and Raveendran K. Marine mycoflora of Gujarat (India). *Bioinfolet.* 2012, 9, 289-291.

Gayatri, R. Nambiar and Raveendran K. *The mangroves and the manglicolous marine fungi.* Malabar Natural History Society, Calicut, Kerala, India, 2013a.

Gayatri, R. Nambiar and Raveendran K. Marine mycoflora of Karnataka. *Geobios.*2013b, 40, 87-91.

Gayatri, R. Nambiar and Raveendran K. Filamentous marine fungi on *Acrosticum* and *Salvinia. Bioinfolet.*2013c, 10, 839-840.

Gayatri, R. Nambiar and Raveendran K. Marine mycoflora on *Avicennia*sp and *Kandeliacandel* along Kerala coast (India). *Geobios.*2014, 41, 84-90.

Hyde, K.D. and Jones, E.B.G. Marine mangrove fungi. *Marine Ecol.*, 1988, 9, 15-33.

Kohlmeye, J. and Kohlmeyer, E. *Marine mycology – the higher fungi.* Academic Press, New York, USA, 1979.

Kohlmeyer, J. and Volkmann Kohlmeyer. Illustrated key to the filamentous higher marine fungi. *Bot. Mar.*, 1991, 34, 1 – 61.

Maria, G.L., and Sridhar, K.R. Richness and diversity of filamentous fungi on wood litter of mangroves along the west cost of India. *Curr. Sci.*, 2002, 3, 1573 - 1580.

Maria, G.L and Sridhar, K. R. Mangrove filamentous fungi – recent advances. In: *Microbial diversity opportunities and challenges* (eds., Gautam, S. P., Sharma, A., Sandhu, S. and Pandey, A. K.) Shree Publishers and Distributors, New Delhi, India.2005, pp. 241-278

Maria, G.L and Sridhar, K. R. Decomposition of dead twigs of *Avicennia officinalis* and *Rhizophora mucronata* in mangrove in south western India. *Bot. Mar.*, 2006a, 49, 450- 455.

Maria, G.L and Sridhar, K. R. Fungal diversity on mangrove woody litter *Rhizophora mucronata* (Rhizophoraceae). *Indian J. Marine Sci.*, 2006b, 35, 318- 325.

Newel, S.Y. Established and potential impacts of eukaryotic mycelia decomposer in marine/terrestrial ecotones. *J. Experimental Marine Biol.* and *Ecol.*, 1996, 200, 187-206.

Raveendran, K. and Manimohan, P. *Marine fungi of Kerala – A preliminary floristic and ecological study*, Malabar National History Society, Calicut, Kerala, India, 2007.

Ravikumar, D.R. and Vittal, B. P. R. Studies on marine fungi of India. I- Notes on some Ascomycetes. *Kavaka*. 1987, 15, 99-103.

Sarma, V.V. and Vittal, B. P. R. Biodiversity of manglicolous fungi on selected plants in the Godavari and Krishna deltas, east coast of India. *Fungal Diversity*.2000, 6, 115-130.

Sarma, V.V., Hyde, K.D. and Vittal, B. P. R. Frequency of occurrence of mangrove fungi from the east coast of India. *Hydrobiol.*, 2001, 455, 41-53.

Chapter 4

Impact of Urbanization in Kengeri Region of Bangalore Urban: A Case Study

K. Harisha

Department of Environmental Sciences,
Bangalore University, Bangalore, Karnataka
e-mail: harisha_akp@yahoo.co.in

Recently urbanization is a big problem in cities across the world due to increase in the growth of human population. Land use planning is all about ensuring the use of land resource in a systematic way to meet the needs of present and future generations can be best addressed. migration of people from outside areas especially from rural areas to urban is the major cause of urbanisation which leads to increase in the number of urban fringes in the outskirts of the city. This approach is a big threat to biodiversity. Rural areas are urbanised because of less economy and less dependent upon agricultural activities. Urban developments,agricultural activities and industries have grown up along the corridors of water bodies in the Bangalore city. The domestic sewage, industrial effluent, and agricultural waste enter the water bodies causing the severe water pollution. Because of this monitoring programme are necessary to create awareness among the public about the consequences of present and future threats of contamination and encroachment of water bodies leading to the loss of flora and fauna. The present study made in the Kengeri region of Bangalore district showing the dynamics of the LULC classes will help in assessing the health of the ecosystems. Like other areas, urban monitoring and management is also one of application areas of the information sciences and technologies, i.e. GIS and Remote Sensing. These technologies can provide computer aided tools for mapping, monitoring, and analyzing urban dynamics to incorporate the acquired information for management purposes.

Keywords : *Urbanization, Land use, GIS and Remote sensing, Water bodies.*

Introduction

Land is defined as any part of the earth's surface not covered by a body of water; the part of the earth's surface occupied by continents and islands: Land was sighted from the crow's nest. Land use term usually defined more strictly and refers to the way in which, and the purposes for which, humans employ the land and its resources (William,2000). Land use planning has as its basic purpose to ensure that each area of land will be used so as to provide maximum social benefits, especially including food production, without degradation of the land resource (Nor, 2010). The general land use classifications such as forest, agriculture, urban, residence *etc.*, for each country more or less are almost same but some of the country may have more specific land use type (Anthony, 1994).

Urban change is the result of urbanization process. Urbanization, in general, is known as the process of growth in the proportion of population residing in urban places. In summary, we can say that demography and land use pattern change, industrialization and social transformation occur during the urbanization processes.

According to Pradhan and Perera (2005), 2 billion new residents being added to the cities of the developing world in the next 25 years and industrial development will continue putting pressure on land use and environment in the major city areas. In the developing countries, large cities are growing at a higher rate than those in the rest of the world (Thouret,1999). This means that 21st century will be the century of rapid urbanization (Pradhan, Perera,2005). In recent decades, remotely sensed data have been widely used to provide the land use/cover information such as degradation level of forests and wetlands, rate of urbanization and other human-induced changes (Alrababah and Alhamad,2006). Moreover, the advent of widely available and less expensive Landsat imagery has permitted the development of highly accurate land cover map products (Goet, *et al.*, 2009).

Water is a valuable natural resources that is essential for humans to survive and also the ecosystems health. Since in the past few decades, the increasing of anthropogenic activities especially in urban area has lot of effects to Freshwater bodies (Mohd Hasmadi *et al.*, 2010). And this is the global issues which happening throughout the world and Bangalore also face these problems.

The valuable data can be obtained to provide a basis comparison of water body at different seasons and different years. The monitoring and assessment may be useful for research and policy making purposes. *In situ* measurements and collection of water samples for subsequent laboratory analyses are currently used to evaluate water quality. These measurements are accurate for a point in time and space but do not give either the spatial or temporal view of water quality in wide space. Thus, the technologies such as remote sensing and GIS are very useful as a tool in evaluating and monitoring water quality (Mohd Hasmadi *et al.*, 2010).

The sustainable development is an integral component of human settlement development and gives full consideration to the needs and requirement of achieving economic growth and development, social development, social progress employment opportunities that are in harmony with the environment (UNCHS, 2001. and Climate Change 2001).

Though land use changes are indirect consequence of national economic growth, it is important to evaluate land use changes in the regional and the local context in order to assist in anticipating the impacts associated with change and contribute to an understanding of productive environmental sustainability (Laymon.C, 2003).

Nowadays, GIS and remote sensing became important tools for geospatial solution for natural resources. A GIS allows access to large amounts of information quickly and efficiently and remote sensing provides spatial data about land features. The information are visualize in new ways that reveal relationships, patterns, and trends not visible with other popular systems. With the support of satellite image from remote sensing technology, the analysis of geo-morphology of earth surface can be more significant (M.K. Amirin *et al.*, 2010).

Urbanization of Bangalore District

Bangalore has been substantially affected by globalization and rapid urbanization over the last decade. The demand on services and the quality of life in the city is not confined only to the central core or the Bruhat Bangalore Mahanagar Palike jurisdiction but spreads beyond the peri-urban areas, the Metropolitan area and outwards, into the Metropolitan Region (BMR). A commonly accepted fact about the evolution of a city is that migration plays an important role in the growth of its population, however in case of Bangalore: over 60 per cent of its growth comes from natural growth. The migrant population constitutes about 25 per cent. The large proportions of the migrants are well educated and qualified.

Land Use/Land Cover Mapping

The term Land cover relates to the type of feature present on the earth surface of the earth. Corn fields, lakes, concrete highways are all the example of land cover types. In order to use land optimally, it is necessary to have information on existing land cover/land use, and the capability to monitor the dynamics of land use resulting out of newer demands of increasing population and changed life styles. Some of the areas in which land cover/land use maps are important include preparation of development plans in rural and urban sectors, environmental impact assessment, environmental hazard zonation, utility and infrastructure planning, *etc.* (George Joseph, 2008) The classification systems adopted for land cover/land use mapping at different levels of classification are hierarchical in nature. Land use/land cover mapping at National level on a 1:250,000 scale has been carried out for the entire country using satellite data Ajai (2002) based on the number of studies, an indicative guide line for resolution, scale of mapping and its usage is given in the Table 4.2.

Table 4.1: Spatial Resolution, Scale and Area Coverage for Thematic Mapping

Sl.No.	Spatial Resolution	Scale	Area coverage
1	200m-1km	1:1-5m	Regional
2	70-100m	1:250,000	National/State
3	20-40m	1:50000	District
4	10-15m	1:25000	Tehsil/Block
5	1-5m	1:4000-10000	Village/Implementation level

Source: Ajai (2002).

Table 4.2: Land Use/Land Cover Classification System (NRSA)

Sl.No	LEVEL-I	LEVEL-II
1	Built-up land	1.1 Built-up land
2	Agricultural land	2.1 Crop land (a) Kharif (b) Rabi (c) Kharif + Rabi 2.2 Fallow land 2.3 Plantations
3	Forest land	3.1 Evergreen/semi-evergreen forest 3.2 Deciduous forest 3.3 Degraded or scrubland 3.4 Forest blank 3.5 Forest plantations 3.6 Mangroove
4	Water bodies	4.1 River/stream 4.2 Lake/reservoir/tank/canal
5	Waste lands	5.1 Salt affected land 5.2 Water logged land 5.3 Marshy/Swampy land 5.4 Land with or without scrub 5.5 Gullied/ravinous land 5.6 Sandy area
6	Others	6.1 Shifting cultivation 6.2 Grassland/grazing land 6.3 Snow covered/glacial area

Source: George Joseph, 2008.

Land use/Land cover map is prepared by visual interpretation technique using satellite data with the help of Survey of India Topographic maps on 1:50000 scale. LEVEL-II classification of National Remote sensing Agency (NRSA) has been followed for the delineation of maps.

Built Up Land

Urban or Built-up Land is comprised of areas of intensive use with much of the land covered by structures. Included in this category are cities, towns, villages, strip developments along highways, transportation, power, and communications facilities, and areas such as those occupied by mills, shopping centers, industrial and commercial complexes, and institutions that may, in some instances, be isolated from urban areas.

Agricultural Land

Agricultural Land may be defined broadly as land used primarily for production of food and fiber. On high-altitude imagery, the chief indications of agricultural activity will, be distinctive geometric field and road patterns on the landscape and the traces produced by livestock or mechanized equipment. The Level II categories of Agricultural Land are: Cropland and Pasture; Orchards, Groves, Vineyards, Nurseries, and Ornamental Horticultural Areas; Confined Feeding Operations; and Other Agricultural Land.

Forest Land

Forest Lands have a tree-crown areal density (crown closure percentage) of 10 percent or more, are stocked with trees capable of producing timber or other wood products, and exert an influence on the climate or water regime. Forest Land generally can be identified rather easily on high-altitude imagery, although the boundary between it and other categories of land may be difficult to delineate precisely.

Water Bodies

Water bodies include streams, canals, lakes, and reservoirs. Streams are the surface flowing in channels of their own and large stream flowing through extensive areas and receiving many other streams are called rivers. Lake refers to static water body quality in inland lakes is often described as trophic state (nutritional state). A lake chocked with aquatic weeds or a lake with extreme nuisance algal bloom is called eutrophic lake and lake with very clear water is called Oligotropic Lake. The man made lakes are called reservoirs.

Waste Lands

This category consists of gullied/ravenous lands, barren/rocky/stony/sheet rock area and land with scrubs. Gullies are narrow and deep channels developed as a result of wearing away of soil by running water. Gullies deepen from rills, which are tiny channels a few centimeters deep, formed by the impact of rainfall and wearing action of runoff generated therefrom. The next forms of land which add up to waste lands are barren/rocky/stony/sheet rock area. These are lands characterized by exposed massive rocks, sheet rocks, stony pavements or lands with excessive surface accumulation of stones that render them unsuitable for producing any green biomass.

Others

This category of land-use which includes tree grooves and mixed vegetation. Certain lands have a perennial cover of either snow or ice because of a combination of environmental factors which cause these features to survive the summer melting season.

Role of Remote Sensing and GIS

The need for recent data to conduct detailed spatial analysis means urban planning departments are increasingly turning to satellite for a board picture of urban areas untrammeled by boundaries on the ground. Over the past two decades, data from earth sensing satellites have become important in mapping the earth's features, infrastructure managing, managing natural resources and studying environmental change. Remote sensing and GIS are providing new tools for advanced ecosystem management. The collection of remotely sensed data facilitates the synoptic analysis of earth-system functions, patterns and change at local, regional and global scales over time. Maps help us to measure the extent and distribution of resources, analyze resource interactions and identify suitable locations for specific action (*e.g.* development or preservation) and plan future events. GIS helps in performing varied analysis on the data thus obtained from remote sensing In this research context it would have been merely impossible to quantify the-use change and its transformation in the absence of these technologies. GIS has played an immense role in the present research in terms of quantification of change and spatial analysis of change.

Application of Remote Sensing and GIS

Remote sensing data provides a permanent record of Topographical features of the area, and adding to it is the fourth dimension of time. The profusion of information contained in a photograph is much greater than that afforded by any other cartographic device. The camera or sensor often reveals the ground information that would have been ordinarily missed by human eye.

Data acquired by aerial as well as remote sensing has been accepted as a useful source for mapping and making inventories of natural resources. This data has been successfully used in the fields of land use/land cover mapping, Forest mapping, snow hydrology, reservoir sedimentation, river morphology, watershed conservation, flood estimation, geological investigation, soil mapping, crop yield forecasting, and also remote sensing has been used in many applications related to water resources such as surface water body mapping, ground water targeting, wetland inventory, flood monitoring, water quality monitoring, run-off modeling, snow cover monitoring, glacier inventory, glacial mass balance studies, irrigation water management and many more

Aim and Objectives of the Study

The study attempts to examine changed pattern of the urban land use/land cover in Kengeri regions over a period of fourteen years (14) thus providing explanation to the structural changes and the changed processes in a typical unplanned urban settlement.

- To study the Urban Land Use/Land Cover Changes using Remote sensing and GIS.
- To assess the change in Surface Area of 15 Perennial water bodies using Remote sensing Data and GIS.

Review of Literature

Remotely Sensed Satellite data has been in use for Land use/Land Cover studies is important for planning and Management activities concerned with the surface of the earth from many decades. The use of panchromatic, medium scale aerial photographs to map land use as been an accepted practice since 1940's. More recently, small scale aerial photographs and satellite images have been utilized for Land use/Land Cover mapping (Lillesand and Kiefer, 1999). The information on land use can now be more easily and reliably obtained from remotely sensed and imagery (Adeniyi, P.O, 1981).

Generally remote sensing image (platforms) do not record activity or the way land is being used (*i.e.*, forest, agricultural, residential or industrial) directly but acquires response based on characteristics of land surface. Satellite remote sensing is effective in providing multitemporal and multispectral data along with the required information for understanding and monitoring land development patterns and processes for building land use and land cover datasets (Weng, Q., 2002) A number of researchers had used multispectral remote sensing to evaluate land use/land cover changes at the micro scales and Also examined changes in land use pattern over a period. It is often not sufficient to analyse the changed process in Land use/land cover but the also to understand the forces at play in the changed process (Brandon, R., and B.A. Bottomley, 2000) and (Mass J.F., 1998).

To infer that land cover conversion involves a change from one cover type to another while land cover modification involves alterations of the structure or function without a wholesale change from one type to another (Briassoulis, H., 2000). The land use change is likely to cause land cover change, but land cover may change even if the land use remains unaltered (Turner *et al.,* 2000).

The data obtained from different sensors have different resolutions, hence some sensors have an advantage over others. IRS LISS-II AND Landsat TM data and SPOT data are useful to prepare maps on 1:50,000 scale, depicting the Level-

II classes (Pathan, 1992). And for selection of best sites for urban development and infrastructure development is carried out using multithematic layers in GIS environment (Pathan *et al.,* 1991, 1993).Spatial information on Land use/Land Cover is required at different scales depending upon its use and when digital data is converted to a map, the spatial resolution plays a very important role on the scale of mapping. Based on number of studies (Ajay,2002) has given an indicative guide line for resolution, scale of mapping and its usage.

Remote sensing of change detection is a process of determining and evaluating differences in a variety of surface phenomena over time. For detection of land cover change, multi temporal data of Landsat TM were found to be more suitable for identification of deforestation areas, mapping there generation/regrowth of forest area and tracing major changes in land cover. Although the methodology for visual interpretation of satellite data for vegetation/land cover classification and mapping has been standardized and made operational in India, the potential of high resolution data such as IRS LISS II and LISS III for extracting vegetations. Related information through visual interpretation (Ramachandran *et al.,* 2000).

Various methodologies have been adopted to study land use dynamics. Remotes sensing data, GIS analysis, modeling and socio-economic studies have been used. Visual interpretation of remote sensing data was used to monitor land use/land cover changes in Maripeda mandal of Warangal district in Andhra Pradesh. The qualitative and quantitative aspect of bio-physical resources has to be taken into account when a new land pattern is introduced. These findings provide sufficient evidence for the relevance of using remotely sensed data for change detection process. With the change detection process it is possible to identify the processes and participants in land use changes. The ultimate usefulness of remote sensing lies in the application aspects of gathered information and this technology is behind utilized to study the natural and human systems for their management.

Changes in Land Use in Perak Catchment Zone, Peninsular Malaysia were inferred from the difference between 1989 and 1999 status and land detoriation over the two mapping periods (Mohd Hasmadi *et al.*, 2010).

The changes in the river course of Ganga - Padma river over space and time to delineate the vulnerable zones for environment management were attempted by Hazra and Bhattacharya (1999). In this connection, various geomorphological and geological features have been delineated using visual interpretation techniques.

Space Application Centre (SAC/ISRO) has undertaken the land use mapping for Mumbai, Ahmedabad, and Calcutta and National Remote Sensing Agency (NRSA) for Chennai and Hyderabad (Pathan*et.al,*1989, 1991 and 1992);(NRSA,1994); (Raghavaswamy*et.al*, 1994).

National Remote Sensing Agency has carried out the urban mapping and change analysis for Greater Mumbai (NRSA, 2002, 2005). National Remote Sensing Agency (NRSA), in a joint project with Hyderabad Urban Development Authority, carried out land use and base mapping for the HDA area using Remote Sensing and GIS Techniques (HUDA, 2003).

GIS and remote sensing was used for the development of water resources in Sai-Gad sub-watershed of Almora District, Uttar Pradesh (Mohan *et al.,* 1999). Various thematic maps on the hydrogeomorphological characteristics, elevation, slope, drainage, surface water bodies and land-use have been generated and integrated for the action plan for Water Resources Development. A distributed model with SCS curve number method called as Land Use Change (LuC) model was developed (Mohan and Shresta, 2000) to assess the hydrological changes due to land use modification.

Bangalore Urban District land use mapping and change assessment was carried out jointly by Indian Space Research Organization (ISRO) and Karnataka State Remote Sensing Agency (1996 – 2002) as a proto type study for Natural Resources Census.

Material and Methods

Study Area

Bangalore, the capital of Karnataka, is located in the south-eastern corner of the state. It resembles an inclined rectangle in Outline in shape. Spanning a geographical area of 2,191 sq km, the district lies stretched between the latitudinal parallels of 12°39' to 13°18'N and the longitudinal meridians of 77°22' to 77°52'E. It is surrounded by Bangalore Rural on its North, East, West, and South expect a part of the district on the south and east being bounded by Dharmapuri District of Tamil Nadu State (Gazette, 1990).

Bare rocky outcrops of granites and gneisses raising 30 to 70 meters above ground level are common in the southern portion. The area is much dissected with rapid erosion. The highest point is 908 meters above mean sea level and the lowest point is 720 meters above mean sea level. The hilly terrain in the southern part is covered by small shrubs and bushes.

Quartz deposits are found on the hillock near Kengeri. The purest types of, when crushed or ground to fine powder could be used in many industries such as the manufacture of Glass, pottery and pottery glazes, silica bricks, sand lime and others. Bangalore is considered to be climatically a well favored district situated in the heart of South Deccan of Peninsular India. The highest maximum temperature of 33°C in April to the lowest minimum of 14°C in January. The mean annual rainfall is 859.6mm

The Vrishabhavathi is the only minor river which flows in the taluk and ultimately joins the river Arkavathi. The particulars of sub-basins as follows Arkavathi basin, Ramasandra sub-basin,Valagerahalli, Kenchanahalli *etc.* They are important for urban areas as they provide for a wide range of uses. Bangalore, with annual rainfall of 900 mm (3.0 ft) with three different rainy seasons covering nine months of the year. June to October is the rainy season accounting for 64 per cent of the total annual rainfall in the S-W monsoon period and 324 mm (1.1 ft)during the N–E monsoons (November – December.).

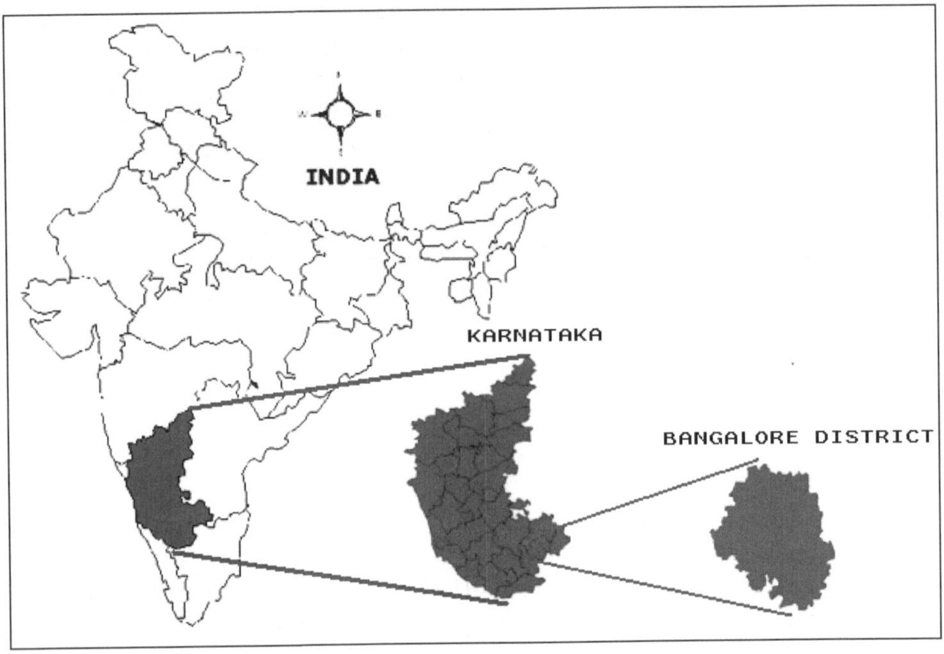

Figure 4.1: Study Area.

The natural vegetation of the district may be broadly grouped into seasonal vegetation, roadside and avenue trees which are planted and cultivated flora. Most of the forest areas are small and are surrounded by agricultural lands, and few species of wild animals are seen and avifauna is rich and varied.

Agriculture is the main occupation of the people on the rural areas of the district and major crops cultivated are paddy, ragi and pulses. And cultivable lands are mainly rain-fed and dry farming is a characteristic feature of the district. The contribution of horticulture to the economy of the district is quite substantial. Animal husbandry has become an important subsidiary occupation due to the predominance of small land holdings.

The Topographical and climatic features of the district are subjected to small regional variations and are favorable for the growth of a variety of plants. Among the timber yielding species, honge, tamarind, nerale, jack fruit, Acacias, ippe and eucalyptus. Trees which are lesser older are useful for firewood are found in the forest areas. Sulikere forest research institute is located in Sulikere of Kengeri hobli.

Manufacturing industry enjoys a pivotal place in the economy of the district. Bangalore is called the Silicon Valley of India because of the large number of information technology companies located in the city. The headquarters of several public sector undertakings such as Bharat Electronics Limited (BEL), Hindustan Aeronautics Limited (HAL), National Aerospace Laboratories (NAL), Bharat Heavy Electricals Limited (BHEL), Bharat Earth Movers Limited (BEML) and Hindustan Machine Tools (HMT) are located in Bangalore. In June 1972 the Indian Space Research Organization (ISRO) was established under the Department of Space and headquartered in the city.

Data Base

Three Landsat imageries of 1990, 2000 and 2004 for the period of 14 years were used for the study. The date gap of all three acquired imageries was about a week of October and November.

Primary Data

Top sheets (57H/5 and 57H/9), are the primary data which were collected from Survey of India and Three Landsat imageries which were downloaded from the respective websites (www.landsat.org, http://glovis.usgs.gov). The Landsat images used in the study were acquired free of distortions from the sources mentioned above. And were processed geometrically and enhanced.

Secondary Data

Demographic data was collected from Statistical Abstract of Karnataka (2000-2001), Directorate of Economics and Statistics, Bangalore. From this we got the population data, Area (sq. km), Density, decadal growth from 1971-2001.

Field Data

The ground truth was done twice in the study to identify the doubtful areas arised in the imagery during preliminary interpretation stage. And photo's were collected during the ground truth are displayed at the end of the report.

Equipment Used

Different equipment's like Top sheet, Imageries, Softwares-erdas imagine-8.5,ArcGIS-3.2, Computers, Scanners *etc.* all this equipment were used on the process of experimentation.

Base Map is prepared by using Survey of India topographic maps on 1:50,000 scale. All the settlements, road networks, railway networks water bodies and forest areas are taken into consideration by comparing the Survey of India topographic maps with that of the satellite image, the size of all the settlements are increased and updated.

Drainage map is prepared by survey of India topographic maps on 1:50,000. All the steams and water bodies/lakes existing in the study area are marked in this map. The Drainage system existing is dendritic.

Figure 4.2: False Imagery of the Study Area of Bangalore District (*Source*: www.landsat.org)

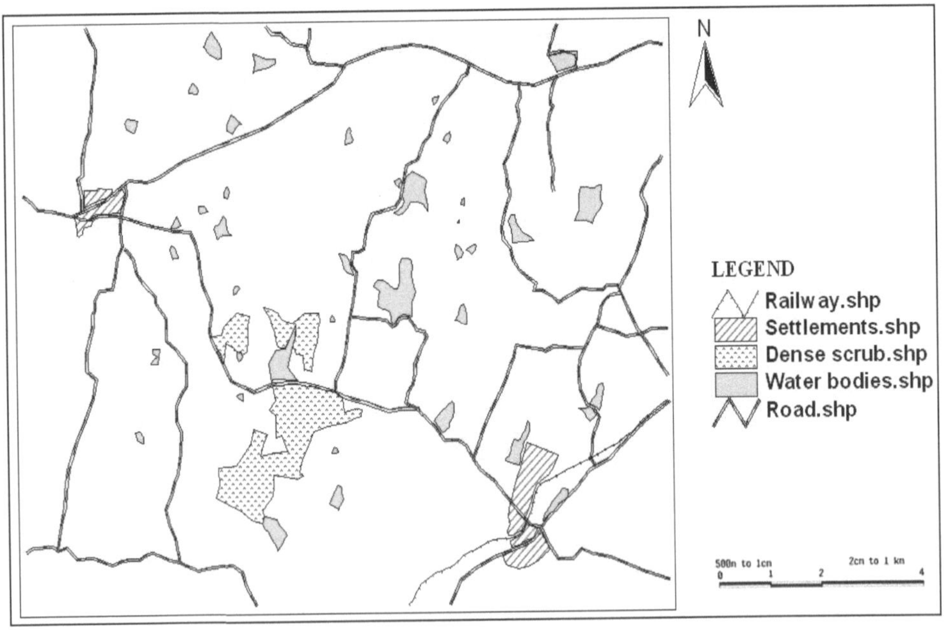

Figure 4.3: Base Map of the Study Area.

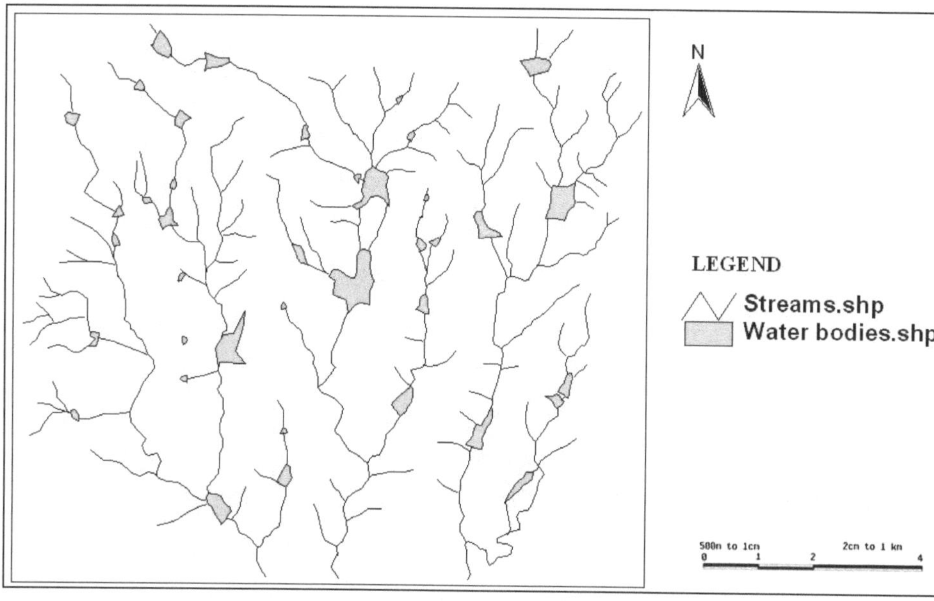

Figure 4.4: Drainage Pattern Map of the Study Area

Satellite Data

Table 4.3: Detail Characteristics of the Landsat Imageries

Sl.No.	*Satellite*	*Sensor*	*Band*	*Imagery Date*	*Path/Row*	*Source*
1	Landsat 5	TM	7	31/10/1990	144/51	Landsat.org
2	Landsat 5	ETM	8	27/11/2000	144/51	Landsat.org
3	Landsat 5	TM	7	22/11/2004	144/51	USGS

Table 4.4: Survey of India's Topographic Maps

Sl.No.	*Topographic Map No*	*Scale*	*Year of Survey*	*Year of Publication*
1	57H/5	1:50,000	1970-71	1973
2	57H/9	1:50,000	1978	1980

Software's Used

The image was processed with various software programs to get required results. The following software programs were used for the study

- ☆ Erdas Imagine 8.5
- ☆ ArcGIS 3.2

Methodology

The methodology adopted to achieve the objectives of the study involves the following steps:

Collection of source data like satellite data of different years, SOI top maps, district/village maps. These are the main inputs for the preparation of thematic layers. Satellite data's acquired free of distortions were geometrically corrected and digitally enhanced. Geo coding and geo-referencing of satellite data by extracting the Ground control Points (GCPs) from the Top sheets.

Preparation of Basic themes like Base map drainage map from the source data. Then updating of base map and drainage map from the satellite image by visual interpretation. Thematic maps like Land Use/Land Cover map are prepared by visual interpretation of the satellite imagery. Visual interpretation is carried out based on the image characteristics like tone, size, shape, pattern, texture, location, association, background *etc.*, in conjunction with existing maps/literature. Preliminary quality check and necessary corrections are carried out for all the maps. Field visits are carried out to check the delineated units of the maps prepared by visual interpretation of satellite imageries. Photographs related to the study area were collected.

Final quality check and necessary corrections are carried out for all the maps prepared. All maps prepared are converted into soft copy by digitization. In that process editing, labeling, mosaicing, quality checking, data integration, are checked out.

Table 4.5: Land Use/Land Cover Area (Estimate) of Kengeri Region (1990, 2000, 2004)

Sl.No.	*Land Use/ Land Cover*	*Area in (Per cent) 1990*	*Area in (Per cent) 2000*	*Area in (Per cent) 2004*
1	Agriculture	89.35	88.27	86.93
2	Built-up	2.50	5.15	7.38
3	Waste lands	4.38	3.04	2.72
4	Water bodies	1.94	2.18	1.84
5	Forest cover	1.83	1.36	1.13
6	Others	0	0	0
	Total	100	100	100

Surface Area of a water body increases/decreases due to seasonal fluctuations like precipitation,temperature, volume of water, emerged and floating vegetation; Urbanization, globalization, population increase and siltation due to soil erosion. Among the 15 perennial water bodies, Bhimanakuppe has large surface area and Gowdanakere has small surface area.

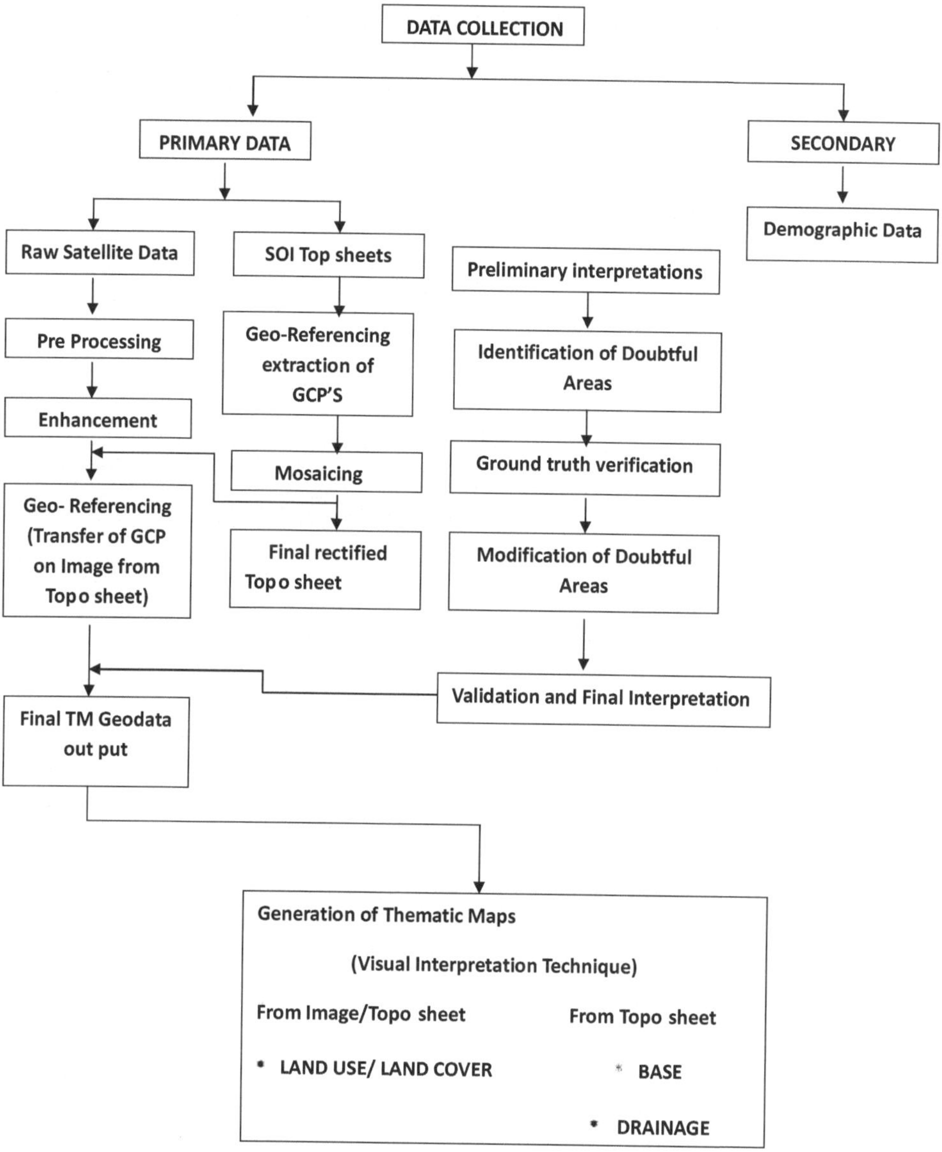

Figure 4.5: Methodology for Generation of Thematic Map.

Conclusion

Remote sensing and GIS is an important tool for measuring the ecosystem dynamics of any particular area. It helps in detecting the type of change, location of the change, quantifying the changes taking place in an ecosystem. Rapid development with population growth, urbanization and land use activities are raising concern over the degradation of the environment. The uses of remote sensing data as input in a GIS environment are necessary to solve land use and

Table 4.6: Water Surface Area of Kengeri Hobli of different Years (1990, 2000, and 2004)

Sl.No.	Water Bodies	1990's Area in (ha)	2000's Area in (ha)	2004's Area in (ha)
1	Bairaiahanakatte	2.46	2.84	2.23
2	Bandimatt (Hoskere)	18.4	18.65	17.75
3	Basavanapalya	2.78	3.63	1.13
4	Bhimanakuppe	45.57	51.63	48.31
5	Gowdanakere	3.34	1.24	1.07
6	Halabhav	4.18	5.55	2.78
7	Herohalli	11.87	12.59	11.31
8	Kodigehalli	2.81	2.69	1.81
9	Kommaghatta	11.43	13.77	10.47
10	Malathalli	22.59	25.43	23
11	Margondanahalli	22.77	24.54	22.67
12	Muddaiahanapalya	10.96	11.67	9.28
13	Ramasandra	41.97	46.41	43.5
14	Ramasandrachikkakere	3.21	2.7	2.61
15	Yellacheguppe	6.64	7.43	6.54

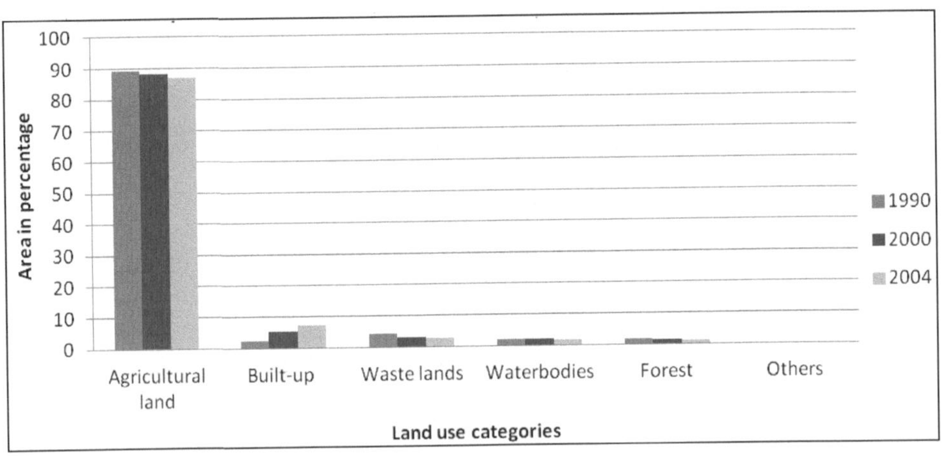

Graph 4.1: Percentage Changes in Land Use Categories in Kengeri Region (1990, 2000 and 2004).

environmental problems. Conversion of agriculture lands, water bodies into increase built-up concrete structures which leads to degradation of the Ecology in the region. And loss in agriculture lands leads to food scarcity. Again that leads

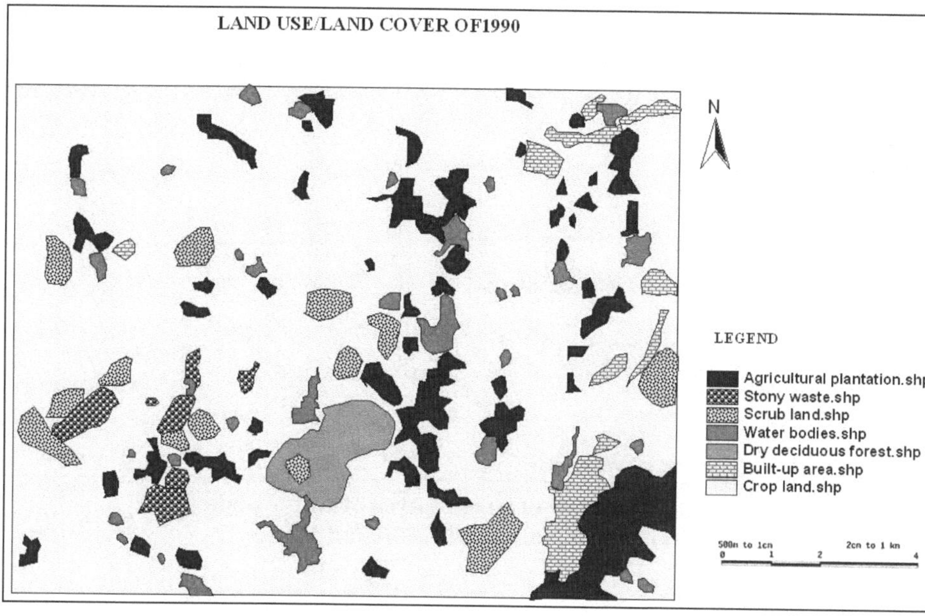

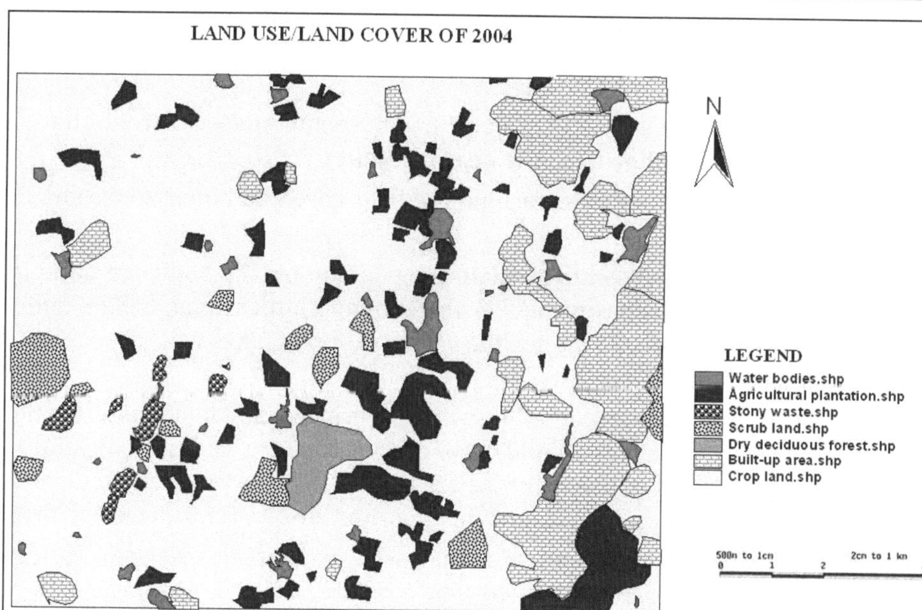

Figure 4.6: Land Use/Land Cover of Bangalore -1990 and 2004.

unsustainable growth in the region. Remote sensing and GIS are most vital tools for land-use studies and It is essential for achieving the sustainable development of the region.

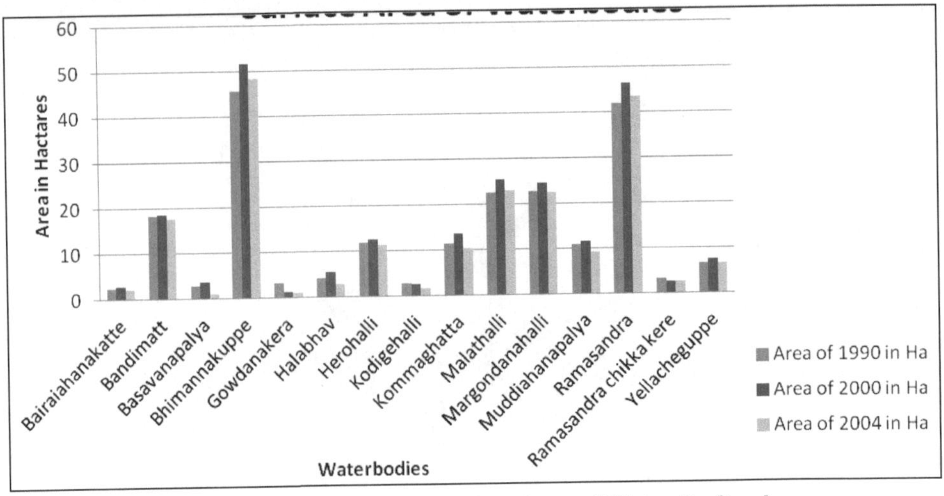

Graph 4.2: Bar Chart of Surface Area of Water Bodies in Kengeri Region (1990,2000 and 2004).

Scope for Future

- ☆ Priority should be given to conserve the catchment area of the Kengeri regions to conserve the lake.
- ☆ Bunds should be raised higher around the water bodies to avoid siltation and contamination from the nearby locality.
- ☆ To restrict the activities around the lake, core and buffer zones should be created.
- ☆ It is in good interest to escalate the position of lake bodies to national parks, so that the control over the core and buffer areas will be under the control of MoEF. Hence the conservation of lakes will take a higher step than development.
- ☆ All around the lake plantations with fast growing and indigenous species of trees and shrubs should be encouraged which will not only add to the grandeur and beauty of the area, but also make the ecosystem more dynamic, providing food, and nesting sites for different species of birds.
- ☆ Most of the people around the lake are dependent on the lake for various purpose such as washing *etc.* hence before planning any development basic needs of the people should be considered.
- ☆ Due to improper planning, sites are allotted such that the houses are built on the bank of the lake, reclamation of the area is recommended.
- ☆ Proper planning of urbanization is suggested around the catchment area so that the rainwater does not get into the sewage and thereby allowing it to flow into the lake.

REFERENCES

Adeniyi, P.O. (1981). Application of aerial photography to the estimation of the characteristics of Residential Buildings. The Nigeria Geographical Journal, 19: 189-200.

Anderson *et al.* (1976). Land use and Land cover Classification System for Use with Remote sensor Data.

Anthony.Y, (1994). Towards an International Classification of Land Use. ConsultancyReport.Food and Agriculture Organization of the United Nations, Rome.

BDA (1995): "Comprehensive Development Plan (Revised), Bangalore", Vol.1 and 2 Bangalore Development Authority.

Brandon. R. and B.A. Bottomley, (2000). Mapping Rural Land Use and Land Cover Change in Carroll County, Arkansas Utilizing Multi-Temporal Landsat Thematic Mapper Satellite Imagery. University of Arkansas.

Briassoulis, H (2000). Analysis of Landuse change: Theoretical and modelling approaches. The Web book of Regional Science. Regional Research Institute, West Virginia University.

DES (2006): Statistical Abstract of Karnataka (2000-2001), Directorate of Economics and Statistics, Bangalore.

George Joseph, (2008).Fundamentals of Remote Sensing, Second Edition, Universities Press (India) Private Limited.

Goetz, S.J. *et., al.* (2009). Application of Multitemporal Landsat Data to Map and Monitor Land Cover and Land Use Change in the Chesapeake Bay Watershed.

HUDA (2003): "A Plan for Sustainable Development: Hyderabad 2020". Draft Master Plan for Hyderabad Metropolitan area, Hyderabad Urban Development Authority, Hyderabad.

Leica Geosystems GIS and Mapping, (2003), ERDAS IMAGINE Tour Guide.

Laymon, C. (2003). Satellite Remote Sensing of Land Use Change. Universities Space Resea M. Lillesand T, and R.W. Keifer, (1999), Remote Sensing and Image Interpretation, Fourth Edition, John Wiley and Sons inc., New York.

Mass, J.F. (1998). Monitoring Land-Cover Changes: A Comparison of Change Detection Techniques. A Paper presented at the Fourth International Conference on Remote Sensing for Marine and Coastal Environments, Orlando.

Mohd Hasmadi Ismail and Norsaliza Usali (2010). Use of Remote Sensing and GIS in Monitoring Water Quality. Journal of Sustainable Development, vol. 3, no. 3, 228-238.

Nor S, (2010). Land Use Planning and Information System: Geographical Information System In Land Use Planning: Directions and Implementation.

NRSA (1989).Manual of Land Use/Land Cover Mapping.

NRSA (1994). "Mapping and Monitoring Urban Sprawl of Hyderabad", Published Project Report.

O. Fabiyi Oluseyi (2006). Urban Land Use Change Analysis of a Traditional City from Remote Sensing Data: The Case of Ibadan Metropolitan Area, Nigeria, A paper published in Humanity and Social Sciences Journal 1 (1): 42-64.

Pradhan. P. and Perera. R. (2005).Urban Growth and Its Impact on Livelihoods of Kathmandu valley, Nepal.

Pathan, S.K. *et., al* (1989): " Urban Land use Mapping and Zoning of Bombay Metropolitan Region using Remote Sensing Data", Photonirvachak, Vol. 17, No. 3.

Raghavaswamy, V. (1994): "Remote Sensing for Urban Planning and Management", in Space Technology and Geography, NRSA, Hyderabad.

Suryanath, U. Kamath (1990). Karnataka State Gazetter-(Bangalore District), Governament of Karnataka Publication.

United Nations Centre for Human Settlements (UNCHS), (2001). and Climate Change (2001): Impacts, Adaptation and Vulnerability.

Weng, Q. (2002). Land use change analysis in the Zhujiang Delta of China using satellite remote sensing, GIS and stochastic modeling. A paper published in Journal of Environmental Management, 64: 273-284.

William, N. (2000). Agricultural and Small Watershed Hydrology: Watershed Characters.

Chapter 5

Biodiversity and its Conservation Approaches

Iqra Rafiq, Z.I. Buhroo and K.A. Sahaf*

College of Temperate Sericulture-Mirgund, Sher-e-Kashmir University of Agricultural Sciences and Technology of Kashmir, Srinagar – 190 025, J&K

**Corresponding author e-mail: iqrabadi88@gmail.com*

Human pressure threatens many species and ecosystems, so conservation efforts necessarily prioritize saving them. However, conservation should clearly be proactive wherever possible. In this article, we assess the biodiversity conservation value, Biodiversity means the variability among living organisms from all sources, including terrestrial marine and other aquatic ecosystems and the ecological complexes of which they are a part; this includes diversity within species, between species and of ecosystems. The conservation of biodiversity constitutes an essential aspect of sustainable development worldwide. The consequences of biodiversity loss will be wide ranging and affect the overall sustainability of ecosystems. It will disrupt food chains and food webs leading to disturbances of ecosystems.

Keywords: *Approaches, Biodiversity, Conservation.*

Introduction

Biodiversity–the variety of genes, species, and ecosystems that constitute life on Earth–provides numerous essential services to society. These include material goods (for example, food, timber, medicines, and fiber), underpinning functions (flood control, climate regulation, and nutrient cycling), and non-material benefits such as recreation. Biodiversity can contribute to agriculture through pollination and pest control, provide carbon storage and sequestration, and positively affect human physical and mental health. Biodiversity also secures long-term flows of benefits from nature by providing resilience to disturbance and environmental change. These and other economic and social contributions are substantial, with recent estimates claiming that the economic value of benefits from bio-diverse natural ecosystems may be 10 to 100 times the cost of maintaining them.

Throughout the world, increasing internet is being expressed in environmental issues, largely as a result of the serious concern that is felt about the present state of both the local and global environment and the predictions that have made about future trends. The awareness is based on a realization that the state of the earth's biological systems is of fundamental importance for human society and that our influence on these systems is increasing exponentially. During the last decade, much of this concern has focused on the issue of biodiversity. Much of the biodiversity is being last through extinction caused by destruction of more and more natural habitats.

Importance of Biodiversity: The Need for Conservation

Biodiversity provides the basic natural resource that sustains the human race. It is the most important national asset and constitutes an enduring resource for supporting the continued existence of human societies. It is biodiversity that basically determines the structure and function of all ecosystems. Wild species and relatives of crop plants contain valuable genes that are of immense genetic value in crop improvement programs. It is assumed that these genetic resources will play a positive and unique role in the development of new cultivars, including restructuring the existing ones that are handicapped due to poor expression or lack in one or the other attribute. Both modern allopathic as well as traditional systems of Ayurveda are heavily dependent on plant-based drugs to treat a wide variety of diseases. A large proportion of rural and tribal population still depends on wild animals and plants for the daily food, medicine, housing and household materials, fodder and fuel wood requirements. Tropical forests also provide raw materials used in industries, help in water conservation and affect the climate. It has been found that rapid deforestation has led to soil erosion, flooding, increasing scarcity of wood for fuel, *etc.* Ecological processes and genetic diversity is essential for the maintenance and sustainability of living resources involved.

Loss of Biodiversity: Causes

Biodiversity both at a species and generic level is shrinking at an alarming rate throughout the world, mostly due to man-made causes. The present day drastic changes in the environment and habitat due to population explosion and unmanaged developmental activities are so unnatural that the species are not getting full liberty of time and space for their survival and adaptive radiation, resulting in loss of biodiversity, which is a global crisis.

The major causes of biodiversity losses are as follows:

(a) Developmental Pressure

(i) Construction
(ii) Forest based industries
(iii) Oil drilling
(iv) Pollution
(v) Resource extraction, *etc.*

(b) Encroachment

(i) Agriculture
(ii) Fishery
(iii) Habitat depletion/change
(iv) Horticulture
(v) Monoculture forestry
(vi) Shifting cultivation, *etc.*

(c) Exploitation

(i) Collection made by research/educational institutes
(ii) Exploitation by local authorities
(iii) Firewood collection
(iv) Food gathering
(v) Poaching
(vi) Unregulated collection of medicinal plants, *etc.*

(d) Human-induced Disasters

(i) Floods
(ii) Oil spillage
(iii) Wildlife depredation
(iv) Forest fires, *etc.*

Conservation Approaches

Conservation paradigms, practices, and policies have shifted over time and have been variably successful (13). In recent decades, traditional approaches to conservation such as the creation of national parks have evolved to encompass awareness of the diverse benefits provided by protected areas, the importance of local conservation initiatives and interests in protected area management, and the need to address the opportunity costs of conservation among the rural poor. Ecological restoration, both within and outside protected areas, is being increasingly applied worldwide. Actions for species, such as targeted habitat management, removal of invasives, captive breeding, and reintroduction have yielded notable successes.

The major approaches to biodiversity conservation include:

(a) Conservation of biodiversity through a network of protected areas including national parks, sanctuaries, biosphere reserves, marine reserves, gene banks, wetlands, coral reefs, *etc.*

(b) Protection of domesticated plant and animal species in order to conserve indigenous genetic diversity.

(c) Protection and sustainable use of genetic resources/germplasm through appropriate laws and practices.

(d) By establishing database at various levels to document support for protecting traditional skills and knowledge for conservation.

(e) Multiplication and breeding of threatened species through modern techniques of tissue culture and biotechnology.

(f) Control of over exploitation of species.

(g) Restriction on introduction of exotic species without adequate investigations.

Traditionally, conservation has been polarized into two separate approaches:

(a) *In situ* Conservation

In situ means in the natural, original place or position. *In situ* conservation includes conservation of plants and animals in their native ecosystems or even in manmade ecosystems where they naturally occur. But it applies only to wild flora and fauna and not to the domesticated animals and plants, because conservation is achieved by protection of population in nature. This method of conservation warrantees the continuation of ecological processes and organic evolution. Thus helping the concerned biota to continuously evolve in response to newer environmental changes. *In situ* conservation includes a system of protected areas of different categories, *e.g.* National parks, sanctuaries, national monument, cultural landscapes, biosphere reserves, *etc.*

(b) *Ex situ* Conservation

In situ conservation includes conservation of sample genetic diversity away from field habitats. *Ex situ* conservation using sample populations is done through establishment of gene banks, which include genetic resource centers, zoos, botanical gardens, culture collections, *etc. Ex situ* conservation is the chief mode of conservation of cultivation crops and domesticated animals. There are two modes of *ex situ* conservation:

1. Conventional method by seeds
2. Role of Biotechnology in Biodiversity Conservation

Role of Biotechnology in Biodiversity Conservation

Biotechnology is defined as any technical application that uses biological systems, living organisms or their derivatives, to make or modify products or processes for specific use. Biotechnology has spurred today's new-found urgency. The relationship between biotechnology and biodiversity is multi directional. Firstly, biotechnology provides new tools for assessment of biodiversity. It is therefore playing an important role in identifying the biological resources themselves. Secondly, it gives us new methods and guidelines for conservation of biodiversity. Thirdly, it provides a great enhanced capacity for wise and efficient utilisation of biodiversity both as a genetic resource for production and in the remediation of degraded ecosystems.

There are four main areas of biotechnology which can directly assist in plant conservation programmes.

(a) Molecular markers technology
(b) Molecular diagnostics
(c) Tissue culture technology
(d) Cryopreservation

(a) Molecular Markers Technology

Marker techniques play a key role in enabling the assessment of plant diversity at genetic level. The elucidation of population structures and gene distribution patterns within ecosystems provide information which can be used to support *in situ* conservation programmes. Molecular markers provide a powerful approach for investigating the genetic basis of variation between different varieties and for manipulating complex genomes. The practical applications of DNA-marker technologies include advising on germplasm collecting missions and gene bank design. A molecular knowledge of genetic diversity can greatly assist the decision making process associated with *ex situ* conservation and facilitate germplasm collection and gene-bank management. DNA marker technology also has an important role in the monitoring of genetic stability in conserved germplasm. Molecular markers can be used to confirm that conserved germplasm retains genetic fidelity.

(b) Molecular Diagnostics in Biodiversity Conservation

One of the risks associated with collection of plant germplasm, especially from wild sources, is the inadvertent introduction of diseases and pests along with plants or seeds. Virus and virus-like diseases pose a special challenge as they are often symptomless in infected plants and require specialized tests to determine their presence. Quarantines are the primary strategy for preventing the entry of diseased materials into the country. The conventional method of testing requires years of testing before they can be introduced to the field. Molecular diagnostic techniques help in detecting and eliminating pathogens from plant germplasm faster. The development of molecular markers has introduced a new generation of diagnostic kits which can detect the invading parasite on the basis of its genetic material. Roosien *et al.* (1993) demonstrated that RAPD markers provide a basis for the development of diagnostic assay to differentiate the potato cyst nematode species. PCR-based assay is used for detection of plant viruses like Potato Virus Y. Thus, these diagnostic kits have been used for assessment of the phytosanitary status of the sample species to be conserved in germplasm.

(c) Tissue Culture Technology in Biodiversity Conservation

The major drawback of *in situ* conservation is that it requires large areas of land for maintaining the species sample of our interest. Secondly, it is very difficult to bring together and use the information available in different institutions worldwide. Thirdly, the *in situ* conserved species cannot evade the environmental hazards, resulting in loss of rare/endangered plant varieties. Tissue culture techniques and germplasm banks have been very useful in overcoming these limitations. It requires minimum storage space, plant regeneration can be undertaken at any time of the year, and information on the conserved species can be made accessible worldwide. This method includes:

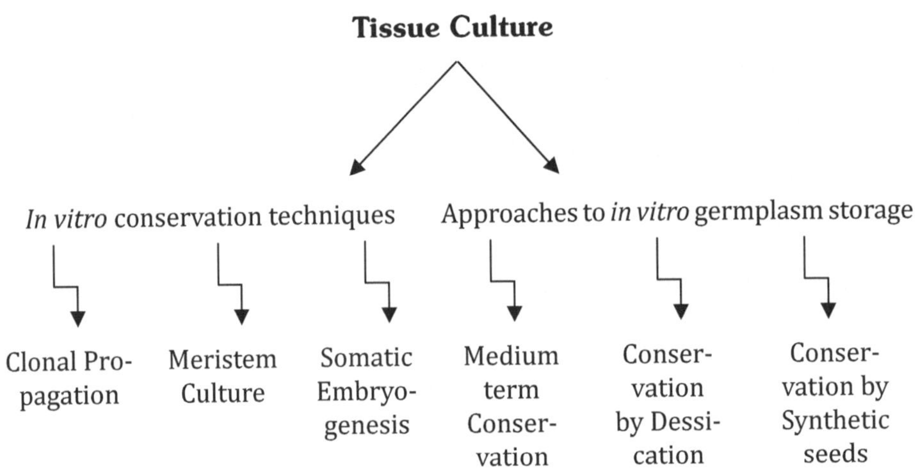

Figure 5.1: Tissue Culture Technology in Biodiversity Conservation

Biotechnological Approaches for Mulberry Germplasm Conservation

Besides the conventional methods of mulberry germplasm collection and maintenance, biotechnological tools have also played an important role in the assessment of genetic diversity of mulberry, mulberry germplasm characterization and its conservation. A few important works related to the application of biotechnological approaches for the conservation of mulberry germplasm are highlighted below:

(i) Molecular Markers Technique

Sharma *et al.* (2000) used fluorescence-based AFLP markers to assess the genetic diversity and establish relationship among mulberry genotypes for conserving the existing genetic resources for future use.

Bhattacharya and Ranade (2001) distinguished nine mulberry varieties using RAPD and DAMD techniques.

Nayak *et al.* (2002) characterized Mysore Local and Victory-1 cultivars through RAPD analysis of 12 collections of each variety from different sources. The analysis suggested a narrow genetic base of the mulberry crop.

Vijayan *et al.* (2004), through RAPD analysis of important Indian mulberry cultivars found that low genetic similarity existed between the group of varieties originating from the Eastern regions and that from Southern regions.

Scientists at CSGRC, Hosur, have made RAPD analysis of 44 mulberry varieties using 5 oligonucleotide primers and revealed 31 discrete well-reproducible amplified products of which 29 were polymorphic. RAPD fingerprinting of 27 Morus laevigata accessions showed polymorphism suggesting genetic variation among the collection.

Isozyme polymorphism using four enzyme systems, namely, anodal and cathodal peroxidase, esterase and acid phosphatase, was studied in 24 Morus laevigata accessions collected through exploration. Grouping on isozyme profiles of anodal peroxidase, cathodal peroxidase and esterase revealed 7, 4 and 5 banding types respectively.

(ii) *In vitro* Techniques

Axillary bud or shoot-tip culture is the most popular method of clonal propagation for mulberry. Bud elongation and shoot development leading to plantlet production was first reported from axillary buds of *Morus alba* cultured on MS medium supplemented with kinctin and NAA (Ohyama, 1970). Thereafter micropropagation procedures have been developed by many workers. Differentiation of embryo-like structures from calli grown on a hormone-free MS medium has been reported (Ohnishi and Kobayashi, 1991). Embryogenesis was reported in mulberry cultures in MS medium supplemented with B5 vitamins and 1 mg/litre of Thidiazurom (Thinh and Katagiri, 1995). Plant regeneration from mulberry protoplasts has not been achieved yet. Artificial seeds of mulberry were produced for the first time by Bapat *et al.* (1987) by encapsulating the axillary buds excised from *in vitro*-raised shoots of *Morus indica* in a sodium alginate matrix. The encapsulated alginate beads were able to sprout on moist filter paper, vermiculite, or agar medium and form plantlets under both *in vivo* and *in vitro* conditions (Bapat and Rao, 1990).

(iii) Storage Techniques

(a) Medium-Term Storage

Morus nigra shoot tips survived for 6 months on multiplication medium at 40 C, but with activated charcoal in the medium, survival could be increased to 42 per cent after 9 months at 25°C, (Wilkins *et al.*, 1988). Sharma and Thorpe (1990) stored 15 genotypes of *Morus alba* L. for six months at 4°C in the dark on shoot-proliferating medium.

(b) Long-Term Storage

The preservation of vegetative buds by freezing them to low temperatures and their sustained regenerative capability were achieved for the first time with *Morus bombycis* (Yakuwa and Oka, 1988). Niino *et al.* (1992) demonstrated that excised shoot tips from winter buds of *Morus bombycis* ev Kenmochi, prefrozen to -20°C at 5°C/day produced more shoots than buds prefrozen at 10°C/day. Partial dehydrating the buds to about 38.5 percent moisture content improved the recovery rates. Alginate-coated winter-hardened shoot-tips of several *Morus* species had maximum shoot formation (81 per cent) when dehydrated to 22-25 percent water content before freezing.

Outlook

Conservation strategies, in concert with other environmental policies, must address seemingly intractable and politically unpalatable issues. In both developed and emerging economies, we need to reduce the carbon and material throughput demanded by current patterns of production and consumption.

REFERENCES

Bapat, V. A., and Rao, P. S. (1990). In vivo growth of encapsulated axillary buds of mulberry (Morus indica L.). *Plant Cell, Tissue and Organ Culture, 20*(1), 69-70.

Bapat, V. A., Mhatre, M., and Rao, P. S. (1987). Propagation of *Morus indica* L.(mulberry) by encapsulated shoot buds. *Plant cell reports, 6*(5), 393-395.

Benson E.E. (1999), Plant Conservation Biotechnology, Taylor and Francis Inc. Padstow, U. K.

Bhattacharya, E., and Ranade, S. A. (2001). Molecular distinction amongst varieties of mulberry using RAPD and DAMD profiles. *BMC Plant Biology, 1*(1), 1-8.

Chopra V.L, Sharma R.P., Swaminathan M.S (1996), Agricultural Biotechnology, Oxford and IBH Publishing Co. Pvt. Ltd.

Heywood V.H., Watson R.T. (1995), Global Biodiversity Assessment, Cambridge University Press, Great Britain.

Kumar U., Sharma A.K. (2003), Plant Biotechnology and Biodiversity Conservation, Agribios, India.

Niino, T., and Sakai, A. (1992). Cryopreservation of alginate-coated in vitro-grown shoot tips of apple, pear and mulberry. *Plant Science, 87*(2), 199-206.

Ohyama, K. A. T. S. U. O. (1970). Tissue culture in mulberry tree. *JARQ, 5*(1), 30-34.

Parihar P. (2008), Environmental Biotechnology, Agribios, India.www.cbd.int.

Roosien, J., Van Zandvoort, P. M., Folkertsma, R. T., Van Der Voort, J. R., Goverse, A., Gommers, F. J., and Bakker, J. (1993). Single juveniles of the potato cyst nematodes Globodera rostochiensis and G. pallida differentiated by randomly amplified polymorphic DNA. *Parasitology, 107*(5), 567-572.

Sharma, A. C., Sharma, R., and Machii, H. (2000). Assessment of genetic diversity in a Morus germplasm collection using fluorescence-based AFLP markers. *Theoretical and Applied Genetics, 101*(7), 1049-1055.

Sharma, K. K., and Thorpe, T. A. (1990). In vitro propagation of mulberry (Morus alba L.) through nodal segments. *Scientia horticulturae, 42*(4), 307-320.

Singh M.P., Singh J.K., Mohanka R., Sah R.B. (2007). Forest Environment and Biodiversity, Daya Publiction House, New Delhi.

THINH, N. T., and KATAGIRI, K. (1995). Induction of direct somatic embryogenesis in mulberry embryos cultured in vitro. *The Journal of Sericultural Science of Japan, 64*(1), 72-74.

Vijayan, K., *et al.*2004 "Molecular evaluation of genetic variability in wild populations of mulberry (*Morus serrata* Roxb.)." *Plant Breeding* 123.6: 568-572.

Vijayan, K., Kar, P. K., Tikader, A., Srivastava, P. P., Awasthi, A. K., Thangavelu, K., and Saratchandra, B. (2004). Molecular evaluation of genetic variability in wild populations of mulberry (Morus serrata Roxb.). *Plant Breeding, 123*(6), 568-572.

YAKUWA, H., and OKA, S. (1988). Plant regeneration through meristem culture from vegetative buds of mulberry (Morus bombycis Koidz.) stored in liquid nitrogen. *Annals of Botany, 62*(1), 79-82.

Chapter 6

Biodiversity of Drosophilidae in Dharwad District

B.S. Srinath* and N. Shivanna

Department of Zoology,
Karnatak University, Dharwad, Karnataka
**Corresponding author e-mail: srinatbs@gmail.com*

The family Drosophilidae is classified under the order Diptera. Among the genera of Drosophilidae, the genus Drosophila represents an extraordinary model system for comparative experimental research. The group has a well-developed phylogeny and an extensive literature on genetics, physiology, neurobiology, behavior, and ecology, molecular, developmental and evolutionary biology. There are a total of 4600 species of Drosophilidae and 1673 species belonging to the genus Drosophila, which is the largest genus reported throughout the world and in India there are 320 species of Drosophilidae and in South India there are 54 species of Drosophilidae and finally in Dharwad district there are 23 species of Drosophilidae reported/ described so far. The present article focuses on the fauna and diversity of Drosophilidae in Dharwad district.

***Keywords**: Taxonomy, Diversity, Drosophilidae, Dharwad.*

Introduction

Class Insecta are the major contributors in the faunal composition of animal fauna throughout the world mainly because of their small size and diverse morphological characters, hence allows them to occupy any ecological niches which cannot be occupied by larger organisms. They are adapted to all the differing environments such as Arctic tundra, Alpine mountains, hot and humid tropical rainforests, coastal mangrove swamps *etc.* They are able to tolerate extreme temperature variations and other environmental conditions. Global species richness of insects is recorded approximately one million (Gullan and Cranston, 2004).

The family Drosophilidae is a large group of acalyptrate flies in the superfamily Ephydroidea. The other large family is Ephydridae, as well as several smaller families, such as Camilidae, Diastatidae, and Curtonotidae (Markow and O' Grady, 2006). Genus *Drosophila* of Drosophilidae family is one such group of flies which has attracted the attention of biologists in terms of taxonomy and Biodiversity studies (Hegde *et al.,* 2001). Several flies of the genus *Drosophila* are used as research models in the studies of genetics, evolution, ecology, development, physiology, behavior, medicine *etc.* Taxonomy studies on *Drosophila* have been continued since the time of Sturtevant (1919), who discovered a new species, *D. simulans* from North America which closely resembled *D. melanogaster*. Sturtevant (1921, 1939, and 1942), Patterson and Mainland (1944) have listed several species of *Drosophila*. Geographical survey of Drosophilids has been made all over the six zoogeographic regions of the world. They are Nearctic species (Wheeler, 1981) which includes the entire continent of North America including Greenland, Neotropical (Val *et al.,* 1981) which includes South America, Central America, the West Indies, Palearctic (Bachli and Rocha Pite, 1981) which includes whole of Europe, North Africa including the north of Sahara, Asia up to the Himalayas, the Japanese Islands, Ethiopian region (Tsacas *et al.,* 1981) which includes Africa, South of the Sahara, Southern Arabia, the Madagascar and adjacent islands, Oriental species (Okada, 1981) which includes India, Ceylon, Tropical China, Japan and the Western Malay Islands, Australia and New Zealand (Bock and Parsons, 1981) which includes Australia, New Zealand, the Eastern Malay Islands, Pacific Oceania (Hardy and Kaneshiro, 1981) which includes groups of Pacific islands including Hawaii Islands. Wheeler (1986) listed about 2240 species of *Drosophila* distributed all over the world. At present in the family Drosophilidae there are 4600 species described consisting of 77 genera. In genus *Drosophila* there are 1673 species among which 792 species belong to subgenus *Drosophila* whereas 344 species belong to subgenus *Sophophora* (Bachli, 2021).

In India, taxonomical studies on *Drosophila* began in 1920s. Bezzi was the first person who first reported *D. repleta* Wollaston from Calcutta (Kolkata). Duda (1923) described a new species, *D. bipectinata* from Darjeeling. *D. prashadi* Brunetti (1923), *D. biarmipes* Malloch (1924) were the other new species described earlier from India. Many new species were later discovered after 1960s from North India, *D. malerkotliana* from Malerkotla, *D. jambulina* and *D. punjabiensis* from Chandigarh (Parshad and Paika, 1964), *D. bambuphila* from Varanasi (Gupta, 1971), *D. orissaensis* from Orissa (Gupta, 1972), *D. kurseongensis* and *D. neoelegans* (Gupta and Singh, 1977), *D. riverata* from Varanasi (Singh and Gupta, 1977), *D. guptai, D. ramamensis* and *D. dominici* from areas near Darjeeling (Dwivedi, 1979, 1981) *etc.* Taxonomical studies on Drosophilidae were reviewed by Gupta (1974) on Indian sub-continent. Kandpal and Singh (2010 a) reported a total of 319 species of Drosophilids from India of which 148 species belong to genus *Drosophila*.

There are new species discovered in some parts of Southern India *i.e., D. rajasekari* from Mysore (Reddy and Krishnamurthy, 1968), *D. mysorensis* from Srirangapattana, Mysore, *D. anomelani* and *D. coonoorensis* from Merupalayam and Coonor, Tamil Nadu; (Reddy and Krishnamurthy, 1970, 1973), *D. s. neonasuta* from Mysore and *D. chamundiensis* from Chamundi hills, Mysore, *D. krishnamurthyi* and *Phorticella striata* from Soundatti and *D. mundagensis* from Mangalore (Sajjan and Reddy, 1975; Sajjan and Krishnamurthy, 1972, 1973, 1975), *D. giriensis* and *D. gundensis* from Bababudangiri and Kemmannugundi hill ranges (Prakash and Reddy, 1977), *D. agumbensis* from Agumbe (Prakash and Reddy, 1978 a), *D. sahyadrii* from Sahyadri hills (Prakash and Reddy, 1979), *D. nagarholensis* from nagarhole region (Prakash and Reddy, 1980), *D. sampagensis* from Sampaje Ghats; *D. bramhagiriensis* and *D. bhagamandalensis* from Bramhagiri and Bhagamandala Ghats; *D. madikerii* and *D. Cauverii* from Coorg district (Muniyappa and Reddy, 1980; Muniyappa *et al.*, 1981, 1982), *D. neoimmigrans* and *D. paraimmigrans* from South Kanara (Gai and Krishnamurthy, 1982, 1986), *D. longivittata* from Salem district (Hegde *et al.*, 1989), *D. palaniensis* and *D. neolacteicornis* from Palani and Nilgiri hills (Hegde *et al.*, 1999).

Apart from the discoveries of new species, various surveys of *Drosophila* fauna have were undertaken to have a detailed account of number of species distributed throughout India such as Punjab (Parshad and Paika, 1964), Srinagar, Pahalgam and Gulmarg of Kashmir valley (Parshad and Duggal, 1965, 1966), Eastern Uttar Pradesh (Gupta, 1969), Bhagalpur (Rahman and Singh, 1969), Andaman and Nicobar islands (Gupta and Ray Chaudhuri, 1970a, b, c; Singh, 1970; Gupta, 1970), Kumaon region which includes districts of Nainital, Almora, Pithoragarh, Bageshwar, Champawat and Udham Singh Nagar (Singh and Bhatt, 1988; Singh and Negi, 1989, 1992, 1995; Singh and Dash, 1993, 1998; Singh and Fartyal, 2002; Singh *et al.*, 2000; Fartyal and Singh, 2000, 2001, 2002; Fartyal *et al.*, 2005; Upadhay and Singh, 2006; Kandpal and Singh, 2010 b), Nagaland (Achumi *et al.*, 2013), Nilgiri, Palani Hills and Kodaikanal range (Reddy and Krishnamurthy, 1971, 1974), Poona and neighbouring areas (Vaidya and Godbole, 1971, 1972, 1973, 1976), Biligirirangana hills (Ranganath and Krishnamurthy, 1972), Charmadi ghat (Gowda and Krishnamurthy, 1972), Bababudangiri and Kemmangundi hill ranges (Prakash and Reddy, 1978 b), Three different localities of Maharashtra state (Hegde and Krishnamurthy, 1980), Dandeli and Ambikanagar (Nagaraj and Krishnamurthy, 1980), Kodagu, Mysore and Dakshina Kannada (Mangalore) districts (Prakash and Ramachandra, 2008). Hegde *et al.* (2001) reviewed biodiversity of *Drosophila* in which they reported 50 species from South India. Guruprasad (2010) assessed diversity of *Drosophila* species from 3 sites in and around Mysore of South Karnataka. *Drosophila* species differ mainly in their levels of genetic diversity, nature and degree of population genetic structure (Markow and O' Grady, 2008).

Among the species found in South India, few are cosmopolitan and most of them are endemic to particular regions and are rarely found. Even though these species are collected they are unable to survive in the laboratory conditions. Thus, such important observation reveals that these flies are sensitive to fluctuations in their habitat, such as light, moisture, humidity, temperature, water *etc.* Barker (1971) and Parsons (1975) studied ecological differences between *Drosophila* species and the total effects of abiotic factors on fitness. Markow and O'Grady (2008) have reviewed feeding and breeding resources of different Species of Drosophilids such as flowers (Brncic, 1983), mushrooms (Jaenike *et al.*, 1983), fruits (Atkinson and Shorrocks, 1977), leaves (Carson, 1971), tree fluxes (Throckmorton, 1975), cactus and soil (Heed, 1977, 1978), land crabs (Carson, 1974) and spider eggs (Hardy, 1965). Some *Drosophila* species are specialists on a single host species (Jones, 2005), on the rotting stems of the cactus (Heed and Kircher, 1965).

In Karnataka state most of the studies on *Drosophila* fauna were concentrated in and around Mysore and the Western Ghats of South Karnataka and North Kanara region (Hegde *et al.*, 2001). No studies reveal any reports or collections from North Karnataka region including Dharwad district. Hence, the present study was mainly undertaken as an initiative to fill this lacuna and understand the diversity of *Drosophila* in Dharwad district.

Materials and Methods

Study Area

Dharwad district is situated in the western sector of the northern half of Karnataka state (Figure 6.1). It lies between the latitudinal parallels of 15° 15' and 15° 35' North and longitudes of 75° 00' and 75° 20' East with an area of 4, 27,329 ha. The district is surrounded by Belgaum district towards North; Gadag district towards East; Haveri district towards South and Uttara Kannada district towards West. The district lies approximately about 800 meters above sea level, which is why it has a moderate and healthy climate.

Dharwad district has two zones according to the physiographical conditions, Dharwad and Kalghatgi taluks were considered under the Sahyadrian transitional belt (semi-malnad) where as Hubli, Kundagol and Navalgund were considered under arid belt (maidan). Flies were collected around all the surrounding areas of these regions and are tabulated in Table 6.1 and their locations are shown in Figure 6.2.

Physiographical Status of Dharwad

The district is classified into two prominent landscapes: Transitional zone/ Semi-Malnad and the Maidan/black soil plains. The semi-malnad features the Sahyadrian region of the western belt of the district. The landscape is characterized by chain of low hill ranges. The hilly landscape of the west passes into the undulating plain of black soil and constitute a transitional belt. Dharwad taluk shares part of its boundaries with Belgaum and with Uttara Kannada districts. It is considered under semi-malnad zone. Kalghatgi taluk shares most of its borders with Uttara Kannada district and is close to the Western Ghats. Hence it is also considered under semi-malnad Zone. The remaining three taluks are considered under the plain lands.

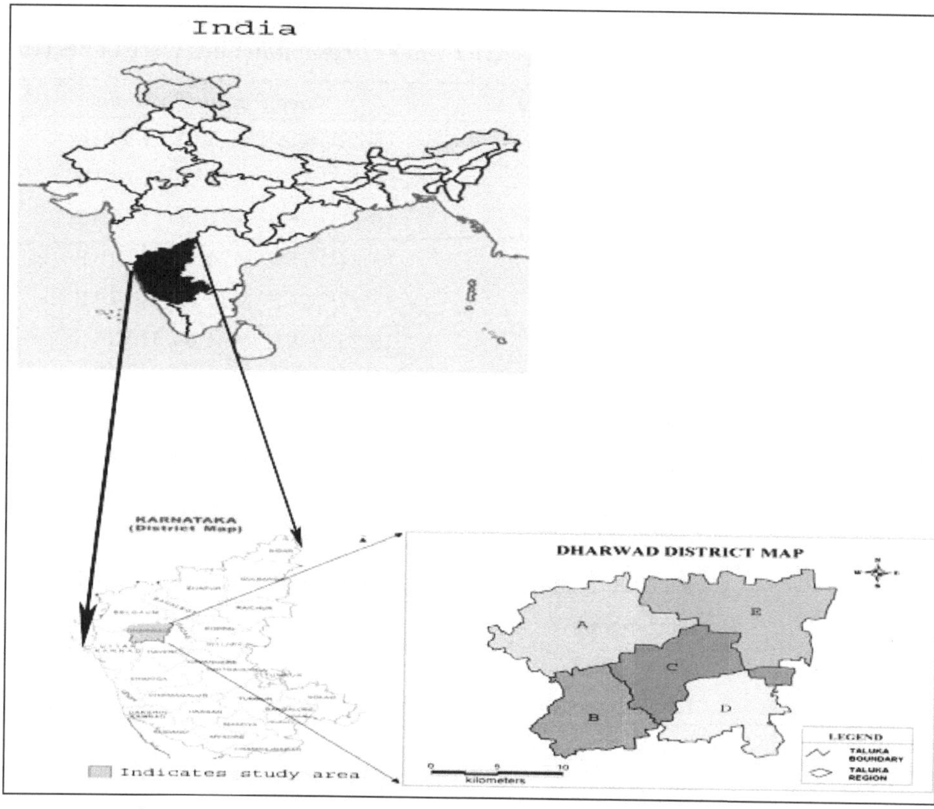

Figure 6.1: Map Showing Geographical Positions of 5 taluks of Dharwad district, Karnataka State in India. A: Dharwad Taluk; B: Kalghatgi Taluk; C: Hubli Taluk; D: Kundagol Taluk; E: Navalgund Taluk.

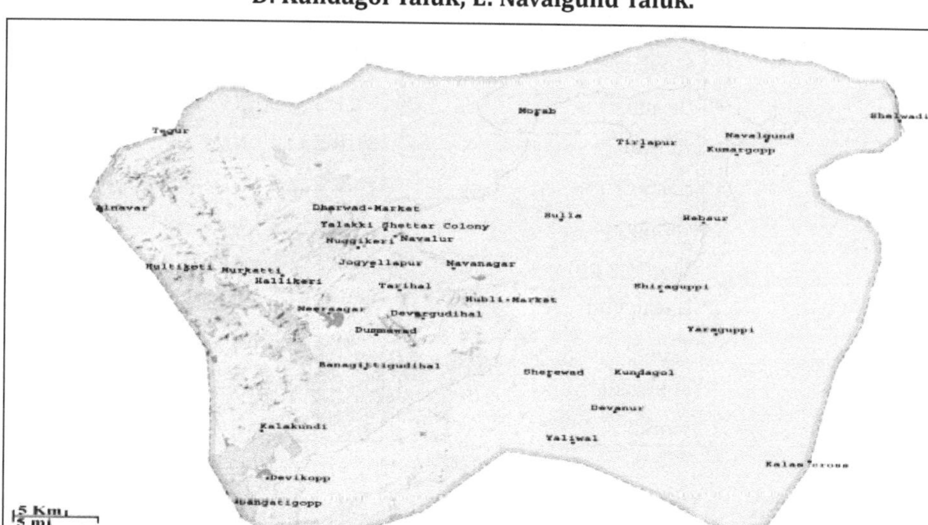

Figure 6.2: Map Showing Localities of *Drosophila* Collection in Dharwad District.

Table 6.1: List of Localities and their Geographic Positions under the Two Physiographical Zones of Dharwad District where *Drosophila* Flies were Collected

	Locality	*Geographic Position*
Dharwad Zone	Navalur	$15^{\circ}25^{1}48.15^{11}$ N $75^{\circ}01^{1}54.69^{11}$ E
(Sahyadrian belt)	KUD campus	$15^{\circ}26^{1}19.57^{11}$ N $74^{\circ}59^{1}01.25^{11}$ E
(Semi-malnad)	Halligeri	$15^{\circ}21^{1}28.61^{11}$ N $74^{\circ}54^{1}17.49^{11}$ E
	Nuggikeri	$15^{\circ}24^{1}15.59^{11}$ N $74^{\circ}59^{1}50.90^{11}$ E
	Jogyellapur	$15^{\circ}23^{1}26.41^{11}$ N $75^{\circ}00^{1}43.87^{11}$ E
	Alnavar	$15^{\circ}25^{1}26.61^{11}$ N $74^{\circ}44^{1}31.58^{11}$ E
	Tegur	$15^{\circ}32^{1}53.27^{11}$ N $74^{\circ}50^{1}15.16^{11}$ E
	Murkatti	$15^{\circ}22^{1}51.13^{11}$ N $74^{\circ}54^{1}02.65^{11}$ E
	Hultikoti	$15^{\circ}21^{1}46.70^{11}$ N $74^{\circ}52^{1}36.23^{11}$ E
	Dummawad	$15^{\circ}19^{1}03.46^{11}$ N $75^{\circ}00^{1}35.81^{11}$ E
	Neersagar	$15^{\circ}19^{1}05.57^{11}$ N $74^{\circ}59^{1}22.20^{11}$ E
	Banagitti Gudihal	$15^{\circ}16^{1}58.72^{11}$ N $75^{\circ}00^{1}37.24^{11}$ E
	Sangatigopp	$15^{\circ}04^{1}44.03^{11}$ N $74^{\circ}53^{1}13.95^{11}$ E
	Kalakundi	$15^{\circ}10^{1}54.43^{11}$ N $74^{\circ}55^{1}04.43^{11}$ E
	Devikoppa	$15^{\circ}07^{1}54.15^{11}$ N $75^{\circ}55^{1}43.09^{11}$ E
Black soil plane (Arid zone)	Tarihal	$15^{\circ}21^{1}19.56^{11}$ N $75^{\circ}02^{1}09.08^{11}$ E
(Maidan)	Sherewad	$15^{\circ}15^{1}23.10^{11}$ N $75^{\circ}10^{1}06.11^{11}$ E
	Shirguppi	$15^{\circ}21^{1}33.07^{11}$ N $75^{\circ}16^{1}52.90^{11}$ E
	Sulla	$15^{\circ}26^{1}54.92^{11}$ N $75^{\circ}10^{1}34.51^{11}$ E
	Devargudihal	$15^{\circ}19^{1}26.08^{11}$ N $75^{\circ}03^{1}29.14^{11}$ E
	Hebsur	$15^{\circ}26^{1}56.34^{11}$ N $75^{\circ}18^{1}12.76^{11}$ E
	Kundagol	$15^{\circ}15^{1}19.58^{11}$ N $75^{\circ}14^{1}50.86^{11}$ E
	Devanur	$15^{\circ}12^{1}56.03^{11}$ N $75^{\circ}13^{1}43.82^{11}$ E
	Yaliwal	$15^{\circ}10^{1}33.23^{11}$ N $75^{\circ}11^{1}27.65^{11}$ E
	Kalas Cross	$15^{\circ}08^{1}22.60^{11}$ N $75^{\circ}25^{1}48.08^{11}$ E
	Yaraguppi	$15^{\circ}18^{1}31.72^{11}$ N $75^{\circ}18^{1}45.97^{11}$ E
	Kumargopp	$15^{\circ}32^{1}25.36^{11}$ N $75^{\circ}20^{1}09.84^{11}$ E
	Navalgund	$15^{\circ}33^{1}33.02^{11}$ N $75^{\circ}21^{1}06.45^{11}$ E
	Tirlapur	$15^{\circ}32^{1}51.85^{11}$ N $75^{\circ}14^{1}48.80^{11}$ E
	Shelwadi	$15^{\circ}35^{1}05.46^{11}$ N $75^{\circ}28^{1}19.72^{11}$ E
	Morab	$15^{\circ}34^{1}58.83^{11}$ N $75^{\circ}09^{1}29.79^{11}$ E

Vegetation and Forest Division

The vegetation of the district includes few patches of moist deciduous forests, dry deciduous forests, scrubby forests and Grasslands. It also includes number of water bodies which inhabit diverse hydrophytes, parasites, semi-parasites, epiphytes *etc.* Moist forests are found in few parts of Dharwad and Kalghatgi taluks. Dry deciduous forests are found in many parts of the Dharwad and Kalghatgi taluks, Scrubby forests are the predominant vegetation type found in the district which inhabits all the five taluks and predominant in Hubli, Kundagol and Navalgund taluks. Grasslands are found commonly in Hubli, Kundagol and Navalgund taluks. It is also found in few parts of Dharwad taluk (Parimala, 2012).

Forest includes three divisions. They are Dharwad division (which includes Dharwad taluk), Hubli division (which includes Hubli, Kundagol and Navalgund) and Kalghatgi division (which includes Kalghatgi taluk). The total forest covers about 35, 711, 988 hectares out of which 35, 754, 081 hectares are considered under the reserve forests and 53, 197 hectares are included under the protected forests. This covers about 8.24 per cent of land under forest. The forest land is more concentrated in Kalghatgi (28.39 per cent), Dharwad (12.23 per cent) and Hubli (2.75 per cent) taluks. There is no forest land in Kundagol and Navalgund taluks. Monsoon deciduous forests are found in the malnad zone which includes Kalghatgi and part of Dharwad taluks (Motebennur, 2012).

Collection of *Drosophila* Flies

Collection was done from 2010 to 2014.

Collection of *Drosophila* flies were done by two methods:

a. Bottle Trapping Method

This method was followed as described by Hegde *et al.* (2001). Approximately 200 ml pint milk bottles were filled with smashed banana fruit about 1 cm and few drops of yeast solution were added. These bottles were tied to tree branches at minimum of 2 ½ feet above the ground level, where maximum shade was available especially during summer and monsoon seasons. Flies were allowed to enter the bottle for a period of 48 hours and the bottles were plugged using cotton. Maximum of 8 bottles was used per site. This was done in order to collect maximum number of flies in a single collection.

b. Net Sweeping Method

Banana fruit smash was prepared and yeast was mixed proportionately with it in order to initiate fermentation faster. The fruit smash thus prepared was spread in the field having sufficient shade from direct sunlight, moisture in the soil and no animal's entry to the vicinity of the locality. After 48 hours, flies were collected by sweeping with insect collecting net over the fruits.

Categorization and Identification of Species

After the collection, flies were brought to the laboratory and were etherized using Diethyl ether and separated into their respective sexes based on their characters such as shape, pattern and number of the abdominal segments and the genital plate under the stereo-microscope. The number of individual males and females were recorded and the females were placed individually into vials containing wheat cream agar medium as prepared following the procedure of Shivanna *et al.* (1996). Since it is difficult to identify the females due to their similar morphology, they were allowed to breed. After the emergence of flies the male progeny was observed for species identification.

Preparation of Wheat Cream Agar Medium

Wheat cream agar medium was prepared according to the procedure described by Shivanna *et al.* (1996). One litre of distilled water was boiled in a vessel (Cooker), 100 grams of Jaggari (Sugar molasses) was dissolved in it. Then 100 grams of wheat cream was added and it was boiled thoroughly. 10 grams of Agar agar and 7.5 ml of propionic acid was added and allowed to boil for 10 – 15 minutes. After boiling, the hot media was poured into the 100 ml culture bottles. After cooling of media, the water vapours on the sides of the culture bottles were blotted off with the help of blotting paper. Yeast solution was added into the bottles and it was plugged with sterilized cotton, the bottles were used for the culture of *Drosophila.*

Identification of Flies

For identification of species the keys described by Hsu (1949); Okada (1956); Parshad and Paika (1964); Bock and Wheeler (1972); Markow and O' Grady (2006) was followed. The male flies collected by above methods were directly used for identification of species. The most important morphological characters employed to categorize species were as follows

Size of the Imagos (in mm)

Head

Number and nature of aristal branches; presence or absence and arrangement of proclinate and reclinate bristles; colour of the eye; shape of gena and nature of carina

Thorax

Ratio of sterno – pleural bristles, number of acrostichal hairs, presence of dorsocentral and scutellar bristles

Wing

Length of the wing and its indices as mentioned by Okada (1956)

Legs

The presence or absence of sex comb on the foreleg of males was observed, if present their pattern such as transverse, longitudinal or oblique and number of sex comb rows and number of teeth in each row. Species belonging to subgenus *Drosophila*, Genus *Scaptodrosophila*, Genus *Zaprionus* and Genus *Phorticella* do not have sex comb hence in such species the shape of the genital plate was observed.

Abdomen

The colour of abdomen whether dark or light was observed

Genital Plate

The structure of genital plate was also observed. The main characteristics studied are the shape of genital arch; shape of anal plate; presence or absence of toe, heel; number of bristles present on the toe region; number of teeth present in the primary surstylus; presence or absence of secondary surstylus, shape and structure of the phallic organ was also studied.

Mounting of the Genital Plate

The genital plate was mounted following the procedure mentioned by Hsu (1949); Parshad and Paika (1964). The posterior portion of the male's abdomen was dissected under the dissecting microscope. The tip of the abdomens were boiled in 10 per cent KOH for about 20 minutes and washed in tap water. The dissection was carried out in a drop of pure glycerine. Later the genital plates were dehydrated in 90 per cent ethanol, absolute alcohol and finally mounted in a mounting medium. Later the slides were observed under the microscope.

Diversity Analysis

Diversity of the flies of different localities of 5 taluks was calculated for Simpson Index (1-D), Shannon index (H) and Equitability index (J), using PAST 3.X software. Simpson Index is performed to analyze the probability that two individuals randomly selected from a sample belong to the same species. Formula to calculate Simpson index is $D = 1 - \Sigma n_i (n_i - 1)/N (N-1)$ where n_i = the total number of organisms of particular species; N= the total number of organisms of all population. Shannon index can be calculated using the formula $H = -\ln \Sigma p_i \ln p_i$ where p_i = proportion of individuals belonging to the same species in a given sample. The equitability index was also calculated using the formula $J^{|} = - \ln \Sigma p_i \ln p_i / \ln N$. This is the Shannon Weiner diversity Index divided by ln N. This index measures the uniformity or equity of the species distribution (Magurran, 2004).

Results

The list of *Drosophila* species collected from different localities was given in Table 6.2. A total of 23 species were obtained from different localities of Dharwad district. Among 23 species collected, 18 belonged to Genus *Drosophila*; 2 belonged

Table 6.2: List of *Drosophila* Species Recorded from Dharwad District

Genus *Drosophila* Fallen 1823
Subgenus *Sophophora* Sturtevant 1939
***melanogaster* species group**
***melanogaster* Subgroup**
1. *Drosophila melanogaster#* Meigen 1830
***ananassae* Subgroup**
2. *D. ananassae#* Doleschall 1858
3. *D. malerkotliana* Parshad and Paika 1964
4. *D. bipectinata* Duda 1923
***suzukii* Subgroup**
5. *D. rajasekari* Reddy and Krishnamurthy 1968
***eugracilis* Subgroup**
6. *D. eugracilis* Bock and Wheeler 1972
***montium* Subgroup**
7. *D. kikkawai* Burla 1954
8. *D. jambulina* Parshad and Paika 1964
9. *D. punjabiensis* Parshad and Paika 1964
10. *D. bhagamandalensis* Muniyappa, Reddy and Krishnamurthy 1981
***takahashii* Subgroup**
11. *D. takahashii* Sturtevant 1927
Subgenus *Drosophila* Sturtevant 1939
***immigrans* species group**
***immigrans* subgroup**
12. *D. immigrans* Sturtevant 1921
***nasuta* subgroup**
13. *D nasuta nasuta* Lamb 1914
14. *D sulfurigaster neonasuta* Sajjan and Krishnamurthy 1973
***repleta* species group**
15. *D. repleta#* Wollaston 1858
***polychaeta* species group**
16. *D. Polychaeta** Patterson and Wheeler 1942
17. *D. daruma** Okada 1956
18. *D. latifshahi** Gupta and Raychaudhuri 1970
Genus *Scaptodrosophila* Grimaldi 1990
19. *Scaptodrosophila nigra* Grimshaw 1901
20. *S. krishnamurthyii* Sajjan and Reddy 1975

Genus *Zaprionus* Coquillet 1902
Subgenus *Anaprionus*
21. *Zaprionus bogoriensis* Mainx 1958
Genus *Phorticella* Duda 1923
Subgenus *Xenophorticella* Duda 1929
22. *Phorticella striata* Sajjan and Krishnamurthy 1975
Genus *Chymomyza* Cznery 1903
23. *Chymomyza vaidyai** Okada 1976

* First time reported in South India; #: Cosmopolitan species.

to Genus *Scaptodrosophila*; 1 each belonged to Genus *Zaprionus*, Genus *Phorticella* and Genus *Chymomyza*. Among 18 species of genus *Drosophila*, 11 species belonged to subgenus *Sophophora* while 7 species belonged to subgenus *Drosophila*. All the 11 species collected under subgenus *Sophophora*, belonged to *melanogaster* species group. Among *melanogaster* species group, all the 11 species belong to 6 different subgroups such as *melanogaster, ananassae, suzukii, eugracilis, montium* and *takahashii*. Among 7 species of subgenus *Drosophila*, 3 species belonged to *immigrans* species group and 3 belonged to *polychaeta* species group while one species belonged to *repleta* species group. Among the *immigrans* species *D. immigrans* belonged to *immigrans* subgroup whereas other two species belonged to *nasuta* subgroup.

Diversity Indices Calculation for different Taluks of Dharwad District

Subgenus *Sophophora*

Dharwad and Kalghatgi revealed maximum species richness and minimum species were recorded in Kundagol taluk. For Simpson, Shannon and Equitability index Hubli taluk recorded maximum diversity, whereas Kundagol recorded minimum diversity for Simpson and Shannon; Whereas, Kalghatgi taluk recorded less diversity for Equitability index (Table 6.3).

Subgenus *Drosophila* and other Drosophilids

Table 6.4 showed that Dharwad and Kalghatgi showed maximum species richness whereas Kundagol and Navalgund recorded minimum. Hubli locality recorded maximum diversity for Simpson and Shannon whereas Dharwad and Navalgund recorded minimum for Simpson and Shannon diversity respectively. Lastly Kundagol locality recorded maximum diversity for Equitability index whereas Dharwad taluk recorded minimum (Table 6.4).

Table 6.3: Diversity Indices of *Sophophora* Subgenus different Taluks of Dharwad District

Diversity Indices	*Dharwad*	*Kalghatgi*	*Hubli*	*Kundagol*	*Navalgund*
Species richness	10	10	7	5	6
Simpson index (1-D)	0.7682	0.659	0.7903	0.5941	0.7684
Shannon index (H)	1.704	1.293	1.708	1.099	1.567
Equitability Index (J)	0.7401	0.5616	0.8777	0.5408	0.8745

Table 6.4: Diversity Indices of *Drosophila* Subgenus and other Drosophilids from different Taluks of Dharwad District

Diversity Indices	*Dharwad*	*Kalghatgi*	*Hubli*	*Kundagol*	*Navalgund*
Species richness	9	9	6	4	4
Simpson index (1-D)	0.6109	0.6751	0.7462	0.6603	0.657
Shannon index (H)	1.214	1.244	1.487	1.221	1.145
Equitability Index (J)	0.5524	0.6943	0.8296	0.8806	0.8263

Diversity Indices Calculated Under Semi-Malnad and Maidan Region

Subgenus *Sophophora*

Table 6.5 revealed maximum species richness for semi-malnad region. When diversity indices were analysed it revealed maximum diversity in maidan than semi-malnad region for Simpson index and Equitability index respectively. But maximum diversity in semi-malnad than maidan region for Shannon index was calculated.

Table 6.5: Diversity Indices of *Sophophora* and *Drosophila* subgenus and other Drosophilids subgenus from Semi-Malnad and Maidan zones of Dharwad District

Diversity Indices	*Semi-Malnad*		*Maidan*	
	Sophophora	*Drosophila and other Drosophilids*	*Sophophora*	*Drosophila and other Drosophilids*
Species richness	11	7	10	6
Simpson index (1-D)	0.7226	0.7485	0.7248	0.7863
Shannon index (H)	1.527	1.54	1.504	1.649
Equitability Index (J)	0.6631	0.701	0.7727	0.9201

Subgenus *Drosophila* and other Drosophilids

Semi-malnad recorded maximum species richness when compared to maidan region. But maidan region revealed maximum diversity than Semi-malnad for Simpson, Shannon and Equitability (Table 6.5).

Further Additions to Drosophilidae of Dharwad District from 2017 to 2019

Later in 2017 a rare species was for the first time recorded from Dharwad district. This was also first report from entire Indian subcontinent. It was known as *D. Polychaeta* Patterson and Stone (Srinath and Shivanna, 2017). This species belonged to the *Polychaeta* subgroup of Genus *Drosophila*. Later in 2019 *Chymomyza* vaidyai Okada was also reported from Dharwad district. This was for the first time reported from South India (Srinath and Shivanna, 2019).

Discussion

Drosophila with its ability to occupy varied habitat conditions provides an excellent platform for biodiversity studies. It has been used as an indicator of biodiversity for habitat and habitat disturbance in tropical forest (Linde and Sevenster, 2002). They have shown that diversity indices have no correlation with disturbance. The distribution of *Drosophila* throughout Dharwad district is uneven as some species are widespread and some are endemic. The uneven distribution may be attributed to the differences in ability of the species to adapt to different localities (Parsons, 1973; Hegde *et al.,* 2001). The collections made in different localities have resulted in a total of 23 species of Drosophilids (Table 6.1). Most of them (18) belong to genus *Drosophila* and have been reported earlier from South India and few are reported for the first time. Among these, *melanogaster* species group was abundantly found with 11 species. This is in contention with the view given by Bock and Wheeler (1972) where they suggested the distribution of subgroups like *melanogaster, montium, ananassae* and *takahashii* are more restricted to oriental region which includes Indian subcontinent. The subgenus *Drosophila* is less abundant with only 7 species being recorded, of which *D. n. nasuta* and *D. s. neonasuta* are abundant (data not shown). *D. immigrans* and *D. repleta* were also reported. *D. daruma* and *D. latifshahi* both belonging to *Polychaeta* species group were reported first time. Other Drosophilids include *Scaptodrosophila nigra* and *S. krishnamurthyii* which are rarely found. *Zaprionus bogoriensis* and *Phorticella striata* were also collected which were abundantly found (data not shown).

D. daruma was first described by Okada (1956) from Kyushu Japan. Toda and Peng (1989) reported this species from Guangdong province, Southern China. Chen and Watabe (1993) reported this species from Eastern China. In India *D. daruma* was first reported by Vaidya and Godbole (1976) from Poona and surrounding areas. The distribution of this species in India is still not completely known even though Toda and Peng (1989); Chen and Watabe (1993) have mentioned vaguely that India is one of the regions where *D. daruma* can be found. But India being a vast country with diverse vegetation and climatic conditions provides excellent habitat to *Drosophila* (Gupta, 1974; Hegde *et al.,* 2001) hence it would be important to understand distribution of individual species with respect to different localities. *D. latifshahi* also belonging to *Polychaeta* species group was first described by Gupta and Raychaudri (1970 c) from Chakia forest in North India. They categorized this species under subgenus *Scaptodrosophila*. Later Toda and Peng (1989) reported this species for the first time from Guangdong province, China. They reclassified the taxonomic status of this species and categorized it under *Polychaeta* species group of the subgenus *Drosophila*. The report of *D. daruma* from Kalghatgi forest and *D. latifshahi* from Dharwad can be considered as first report of collection from not only Dharwad district but entire South India (Srinath and Shivanna, 2012, 2014).

S. krishnamurthyii was also collected from few localities of Dharwad region. It is also considered as one of the rare species collected from Dharwad district. It was first described by Sajjan and Reddy (1975) from grape gardens near Soundatti, Belgaum district. This species was not reported in other collection records made in other parts of Karnataka (Prakash and Reddy, 1978 a, b, 1979, 1980; Guruprasad and Hegde, 2006; Prakash and Ramachandra, 2008; Guruprasad *et al.*, 2010). This species is also considered as endemic species (Hegde *et al.*, 2001). *Zaprionus bogoriensis* was also collected from all the localities in Dharwad region. This species was not reported by any of the earlier workers. The description and collection detail was given by Yassin and David (2010) where they have collected this species from Bangalore, Karnataka in the year 2004.

Diversity indices are the tools which help in understanding the most diverse locality both in terms of species richness and abundance. These indices help in understanding the distribution of the species. If only few individuals belong to one species, the diversity will be lower than when all the individuals are equally distributed among species (Magurran, 1988; Linde and Sevenster, 2002). The calculations of diversity indices have shown similar results. The individuality of species of subgenus *Sophophora, Drosophila* and other Drosophilids is more in Kalghatgi and Dharwad taluk when compared to Hubli, Kundagol and Navalgund taluk. This is due to more number of species especially *D. malerkotliana, D. bipectinata, D. n. nasuta* and *P. striata* collected from these regions (Table 6.3).

Simpson index is used when the probability of any two individuals drawn at random from an infinitely large population belonging to the same species. It is heavily weighted towards most abundant species in the sample and is less sensitive to species richness (Magurran, 2004). The diversity of *Sophophora, Drosophila* and other Drosophilids in Hubli taluk was more when compared to Dharwad taluk. The diversity of *Sophophoran* subgenus in Kalghatgi taluk was less even though its species richness was more whereas diversity of Dharwad was less for subgenus *Drosophila* and other Drosophilids (Tables 6.3 and 6.4).

Shannon index calculated for diversity of *Sophophora, Drosophila* and other Drosophilids also revealed same result as Simpson's index where diversity of Hubli was slightly more (Tables 6.3 and 6.4). The Shannon index assumes that individuals of a species are randomly sampled from a large community which is infinite (Pielou, 1975) and all species are represented in the sample (Magurran,

2004). Equitability index of *Sophophora, Drosophila* and other Drosophilids subgenus calculated showed that Hubli and Kundagol had maximum diversity whereas Kalghatgi and Dharwad showed minimum diversity (Tables 6.3 and 6.4). This is because in Kalghatgi and Dharwad, the abundance of single species is more compared to Hubli and Kundagol and the species richness is not considered as significant in determining the diversity.

The two regions semi-malnad and maidan region which are separated on their physiographic conditions revealed maximum species richness for semi-malnad region when compared to maidan region (Table 6.5). This is due to the vegetation and climatic conditions of the locality as few species depend on fruiting of certain plant species (Parsons, 1973). The Simpson index revealed slightly more diversity for *Sophophora, Drosophila* and other Drosophilids to maidan than the semi malnad region that is mainly because of difference in abundance of individual species (Table 6.5). The difference was observed for the Shannon index for subgenus *Sophophora* and subgenus *Drosophila* and other Drosophilids. The Shannon index recorded maximum diversity in semi-malnad for subgenus *Sophophora*; this is because the species richness is also considered ahead of abundance whereas maidan recorded maximum diversity for *Drosophila* and other Drosophilids. The Equitability index for *Sophophora, Drosophila* and other Drosophilids showed more diversity for maidan region than semi malnad (Table 6.5).

The diversity indices have clearly shown a relation with the species composition, its abundance and its application to evaluate diversity of any locality irrespective of its physiographic, climatic, vegetation or seasons has proved to be a major tool in such analysis (Magurran, 2004; Linde and Sevenster, 2002). The present study reveals that irrespective of species richness and abundance, uniform distribution of species is also an important factor. If there is dominance of only one or two species it does not reflect a healthy biodiversity. But the contribution of rare species must also be taken into consideration as they increase species richness.

From this study the abundance of the species can be can be determined in their respective localities as most of them are common species found throughout India and Southeast Asia (Bock and Wheeler, 1972). Another important observation made in this study is that the species of subgenus *Sophophora* are distributed among 6 subgroups. 4 species of *montium* subgroup consists of many endemic species described from Western Ghats by earlier workers (Reddy and Krishnamurthy, 1970, 1973; Prakash and Reddy, 1977, 1978 a, b, 1980; Muniyappa and Reddy, 1980; Muniyappa *et al.*, 1981, 1982; Hegde *et al.*, 1999). The *ananassae* subgroup is also important subgroup in the biodiversity of *Sophophora*. Even though only 3 species were collected from this subgroup, their abundance is a major contribution to the biodiversity. *D. ananassae, D. malerkotliana* and *D. bipectinata* were found in almost all the localities and have dominated in few localities (data not shown). The sympatric species *D. malerkotliana* and *D. bipectinata* have been found to be coexisting in almost equal numbers, same ability to adapt to any tropical habitat, and have the capability of interspecific hybridization, even though there are no reports of interspecific hybridization in nature but at times they can be interbred in the laboratory (Bock, 1978). Further the diversity of subgenus *Drosophila* and other Drosophilids is considerably poor with only 10 species. But these subgroups have a lot of potential for increased species richness as it was proved with one species (*D. Polychaeta*) being collected for the first time not only from Dharwad but from entire India and the other three species (*D. daruma, D. latifshahi* and *C. vaidyai*) from South India.

Conclusion

To conclude, the collections made throughout Dharwad district is only a first attempt which has revealed a considerable account on the Drosophilid fauna. But more collections in the future will certainly help to study the species richness, their composition and diversity of unexplored localities in the surrounding regions of Dharwad and finally such studies are useful as further additions to the present knowledge.

Acknowledgment

Authors are thankful to the chairman, Dept. of Zoology for providing necessary facilities.

REFERENCES

Achumi, B., Hegde, S. N., Lal, P. and Yenishetti, S. C. (2013). Altitudinal and seasonal variation in *Drosophila* species on mount Japfu of Nagaland, a sub – Himalayan hilly state of India. *J. Insect Sci.* 13, article 117. Available online, www.insectscience.org/13.117.

Atkinson, W. and Shorrocks, B. (1977). Breeding site specificity in the domestic species of *Drosophila*. *Oecologia*, 29, 223 – 232.

Bachli, G. and Rocha – Pite, M. T. (1981). Drosophilidae of the Palearctic region. In: The Genetics and Biology of *Drosophila* (M. Ashburner, J. N. Thompson and H. L. Carson, eds), Vol. 3a, pp. 169 – 196. Academic press, London.

Bachli, G. (2021). Taxodros: The database on taxonomy of Drosophilidae. URL: http://www.taxodros.unizh.cw. Date of access: 12- April- 2021.

Barker, J. S. F. (1971). Ecological differences and competitive interaction between *Drosophila melanogaster* and *Drosophila simulans* in small laboratory populations. *Oecologia,* 8, 139–156.

Bock, I. R and Wheeler, M. R. (1972). The *Drosophila melanogaster* species group. *Univ. Tex. Publ.* 7213, 1 – 102.

Bock, I. R. (1978). The *bipectinata* complex: A study in interspecific hybridization in the genus *Drosophila* (Insecta: Diptera). *Aus. J. Biol. Sci.* 31, 197 – 208.

Bock, I. R. and Parsons, P. A. (1981). Species of Australia and New Zealand. In: The Genetics and Biology of *Drosophila* (M. Ashburner, J. N. Thompson and H. L. Carson, eds), Vol. 3a, pp. 291 – 308. Academic press, London.

Brncic, D. (1983). Ecology of flower breeding *Drosophila*. In: The Genetics and Biology of *Drosophila* (M. Ashburner, H. L. Carson and J. N. Thompson, eds), Vol. 3b, pp. 333–382. Academic Press, London.

Brunetti, E. (1923). A new Indian Drosophilid fly. *Rec. Ind. Mus. (Calcutta)*, 25, 303 - 304.

Carson, H. L. (1971). The ecology of *Drosophila* breeding sites. Lyon Arborteum Lecture Number 2 (L. Harold ed.), pp. 1–27. The University of Hawaii, Honolulu.

Carson, H. L. (1974). Three flies and three islands: Parallel evolution in *Drosophila*. *Proc. Natl. Acad. Sci.* 71, 3517 - 3521.

Chen, H. and Watabe, H. (1993). The *Drosophila virilis* section (Diptera: Drosophilidae) from Eastern China, with descriptions of two new species. *Jpn. J. Entomol.* 61(2), 313 - 322.

Duda, O. (1923). Die orientalischen und australischen Drosophiliden-Arten (Diptera) des ungarischen National-Museums zu Budapest. *Ann. Mus. Nat. Hung.* 20, 24 - 59.

Dwivedi, Y. N. (1979). Description of two new species of subgenus *Drosophila* (Drosophilidae: Diptera) from Darjeeling, India. *Proc. Anim. Sci.* 88 (4), 299 - 304.

Dwivedi, Y. N. (1981). *Drosophila dominici* sp. nov. from Darjeeling district, India (Diptera: Drosophilidae). *Orient. Insects*, 15 (4), 449 - 452.

Fartyal, R. S. and Singh, B. K. (2000). List of Drosophilid species so far described and recorded from Kumaon region, India. *Dros. Inf. Serv.* 83, 58.

Fartyal, R. S. and Singh, B. K. (2001). List of Drosophilid species so far described and recorded from India. *Dros. Inf. Serv.* 84, 30.

Fartyal, R. S. and Singh, B. K. (2002).Taxonomic accounts of new species of Drosophilidae (Insecta: Diptera) of Kumaon region, India. *Entomon*, 27 (4), 355 - 364.

Fartyal, R. S., Singh, B. K. and Toda, M. J. (2005). Review of the genus *Leucophenga* Mik (Diptera: Drosophilidae) in India, with descriptions of five new species from northern India. *Ent. Sci.* 8, 405 - 417.

Gai, P. G. and Krishnamurthy, N. B. (1982). *Drosophila neoimmigrans*, a new species from South Kanara, India (Diptera: Drosophilidae). *Entomon*, 7 (4), 493 - 497.

Gai, P. G. and Krishnamurthy, N. B. (1986). *Drosophila paraimmigrans*, a new species from South Kanara, India (Diptera: Drosophilidae). *Entomon*, 2 (1), 35 - 40.

Gowda, S. L. and Krishnamurthy, N. B. (1972). Report of *Drosophila* species in Charmadi Ghats. *Dros. Inf. Serv.* 48, 38.

Gullan, P. J. and Cranston, P. S. (2004). The Insects: An Outline of Entomology, 3rd edition, Blackwell Science.

Gupta, J. P. (1969). A new species of *Drosophila* Fallen (Insecta: Diptera: Drosophilidae) from India. *Proc. Zool. Soc.* (*Calcutta*), 22, 53 - 61.

Gupta, J. P (1970). Description of a new species of *Phorticella Zaprionus* (Drosophilidae) from India. *Proc. Indian Natl. Sci. Acad.* 36 B, 62 – 70.

Gupta, J. P. and Ray- Chaudhuri, S. P. (1970 a). Drosophilidae of Chakia forest, Varanasi, India. *Dros. Inf. Serv.* 45, 168.

Gupta, J. P. and Ray- Chaudhuri, S. P. (1970 b). The genus *Drosophila* (Diptera: Drosophilidae) in Andaman and Nicobar Islands, India. *Orient. Insects*, 4, 169 - 175.

Gupta, J. P. and Ray- Chaudhuri, S. P. (1970 c). Some new and unrecorded species of *Drosophila* (Diptera: Drosophilidae) from India. *Proc. Roy. Entomol. Soc. London* (*B*), 39, 57 – 72.

Gupta, J. P. (1971). A new species of *Drosophila* (Scaptodrosophila) from Varanasi, India. *Am. Mid. Nat.* 86 (2), 493 – 496.

Gupta, J. P. (1972). *D. orissaensis* a new species of *Drosophila* from Orissa, India. *Orient. Insects*, 6, 561 – 563.

Gupta, J. P. (1974). The family Drosophilidae in India. *Indian Biologist*, 5, 7 – 30.

Gupta, J. P. and Singh, B. K. (1977). Two new and two unrecorded Indian species of *Drosophila* from Kurseong Darjeeling. *Ent. Month. Mag. Oxford*, 113, 71 – 78.

Guruprasad, B. R. and Hegde, S. N. (2006). Altitudinal and seasonal fluctuation of *Drosophila* fauna of Chamundi hill. *Dros. Inf. Serv.* 89, 10 – 11.

Guruprasad, B. R., Hegde, S. N. and Krishna, M. S. (2010). Seasonal and altitudinal changes in population density of 20 species of *Drosophila* in Chamundi hill. *J. Insect Sci.* 10, article 123. Available online, www. insectscience.org/10.123.

Hardy, D. (1965). Diptera: Cyclorrhapha II, series schizophora, section acalypterae I, family Drosophilidae. In: Insects of Hawaii (E. Zimmerman ed.), pp. 814. University of Hawaii Press, Honolulu.

Hardy, E. and Kaneshiro, K. Y. (1981). Drosophilidae of Pacific Oceania. In: The Genetics and Biology of *Drosophila* (M. Ashburner, J. N. Thompson and H. L. Carson, eds.), Vol. 3a, pp. 309 – 348. Academic press, London.

Heed, W. B. and Kircher, H. W. (1965). Unique sterol in the ecology and nutrition of *Drosophila pachea*. *Science*, 149, 758 – 761.

Heed, W. B. (1977). A new cactus feeding but soil breeding species of *Drosophila* (Diptera: Drosophilidae). *Proc. Ento. Soc. Washington*, 79, 649–654.

Heed, W. B. (1978). Ecology and genetics of Sonoran Desert *Drosophila*. Symposium on Genetics and Ecology: The Interface. (P. F. Brussard ed.), pp. 109–126. Springer-Verlag, Heidelberg.

Hegde, S. N. and Krishnamurthy, N. B. (1980). Studies on the *Drosophila* fauna from the three different localities of Maharashtra State. India. *Dros. Inf. Serv.* 55, 60 - 61.

Hegde, S. N., Naseerulla, M. K. and Jayashankar, M. (1989). *Drosophila longivittata*, a new species of *Hirtodrosophila* from Salem (Tamil Nadu: India). *Entomon,* 14 (3 and 4), 253 - 256.

Hegde, S. N., Vasudev, V. and Krishna, M. S. and Shakuntala, V. (1999). Description of two new species of *montium* subgroup of *Drosophila* (Diptera: Drosophilidae) from South India. *Entomon,* 24 (2), 149 - 156.

Hegde, S. N., Vasudev, V. and Krishna, M. S. (2001). Biodiversity of *Drosophila* of South India. In: Trends in Wildlife Biodiversity Conservation and Management (B. B. Hosetti and M. Venkateswarulu, eds), pp. 55-71. Daya Publishing House, New Delhi.

Hsu, T. C. (1949). The external genital apparatus of male Drosophilidae in relation to systematics. *Univ. Tex. Publ.* 4920, 80 - 142.

Jaenike, J., Grimaldi, D. A., Sluder, A. E. and Greenleaf, A. L. (1983). Alpha-amanitin tolerance in mycophagous *Drosophila. Science,* 221, 165 - 167.

Jones, C. D. (2005). The genetics of adaptation in *Drosophila sechellia. Genetica,* 123, 137 - 145.

Kandpal, M. C. and Singh, B. K. (2010 a). Revised list of Drosophilid species so far described and recorded from India. *Dros. Inf. Serv.* 93, 11 - 20.

Kandpal, M. C. and Singh, B. K. (2010 b). Drosophilid species collected from Nainital and Almora district, Kumaon region, India. *Dros. Inf. Serv.*93, 38 - 40.

Linde, K. and Sevenster, J. G. (2002). *Drosophila* diversity over a disturbance gradient. *Proc. Exp. Appl. Ento.* 13, 51 - 56.

Magurran, A. E. (1988). Ecological Diversity and Its Measurement. Princeton University Press. Princeton, NJ.

Magurran, A. E. (2004). Measuring Biological Diversity. Blackwell Science Publishing Ltd.

Malloch, J. R. (1924). Two Drosophilidae from Coimbatore. *Mem. Dep. Agr. India,* 8, 63 - 65.

Markow, T. A. and O' Grady, P. (2006). *Drosophila,* A guide to species identification and use. Academic press.

Markow, T. A. and O' Grady, P. (2008). Reproductive ecology of *Drosophila. Funct. Ecol.* 22, 747 - 759.

Motebennur, S. S. (2012). Land-Use under Jowar sown area, production and per hectare yield of Dharwad District of Karnataka: A Geographical Analysis. *Global Research Analysis,* 1 (5), 74 - 76.

Muniyappa, N. and Reddy, G. S. (1980). A new species of the *montium* subgroup of genus *Drosophila* (Diptera: Drosophilidae). *Entomon*, 5 (4), 331 - 334.

Muniyappa, N., Reddy, G. S. and Krishnamurthy, N. B. (1981). Two new species of *Drosophila* from India (Diptera: Drosophilidae). *Orient. Insects*, 15 (2), 215 - 220.

Muniyappa, N., Reddy, G. S. and Prakash, H. S. (1982). *Drosophila cauverii* - a new species of *Drosophila* from Coorg District Western Ghats, South India. *Entomon*, 7 (1), 1 - 5.

Nagaraj H. J. and Krishnamurthy, N. B. (1980). *Drosophila* fauna of Dandeli and Ambikanagar. *Dros. Inf. Serv.* 55, 114.

Okada, T. (1956). Systematic study of Drosophilidae and allied families of Japan. Gihodo co. Ltd., Tokyo, Japan.

Okada, T. (1981). Oriental species, including New Guinea. In: The Genetics and Biology of *Drosophila* (M. Ashburner, J. N. Thompson and H. L. Carson, eds.), Vol. 3a, pp. 261 - 290. Academic press, London.

Parimala, R. (2012). Flora of Dharwad District. (Ph. D thesis). Karnatak University, Dharwad, India.

Parshad, R. and Paika, I. J. (1964). Drosophilid survey of India II. Taxonomy and cytology of the subgenus *Sophophora* (*Drosophila*). *Res. Bull. Punjab Univ.* 15, 225 - 252.

Parshad, R. and Duggal, K. K. (1965). Drosophilidae of Kashmir, India. *Dros. Inf. Serv.* 40, 44.

Parshad, R. and Duggal, K. K. (1966). Drosophilid survey of India. III. The Drosophilidae of Kashmir valley. *Res. Bull. Punjab Univ.* 17, 277 - 290.

Parsons, P. A. (1973). Behavioural and Ecological Genetics. Clarendon Press Oxford.

Parsons, P. A. (1975). The comparative evolutionary biology of the sibling species *Drosophila melanogaster* and *Drosophila simulans*. *Q. Rev. Biol.* 50, 151 - 169.

Patterson, J. T. and Mainland, G. B. (1944). The Drosophilidae of Mexico. *Univ. Tex. Publ.* 4313, 7 - 216.

Pielou, E. C. (1975). Ecological Diversity. Wiley InterScience. New York.

Prakash, A. and Ramachandra, N. B. (2008). Distribution of *Drosophila* flies in eight different altitudes of three districts of Karnataka, India. *Dros. Inf. Serv.* 91, 82 - 87.

Prakash, H. S. and Reddy, G. S. (1977). Two new species of *Drosophila* (*melanogaster* species group) (Diptera: Drosophilidae). *Orient. Insects*, 11, 597 - 604.

Prakash, H. S. and Reddy, G. S. (1978 a). *Drosophila agumbensis*, sp. nov. from Karnataka, South India (Diptera: Drosophilidae). *Orient. Insects*, 12, 259 - 263.

Prakash, H. S. and Reddy, G. S. (1978 b). *Drosophila* fauna of Bababudangiri and Kemmangundi hill ranges (Western Ghats). *Entomon,* 3(1), 85 - 90.

Prakash, H. S. and Reddy, G. S. (1979). *Drosophila* fauna of Sahyadri hills (Western Ghats) with description of a new species. *Proc. Indian Acad. Sci.* 88B part I (1), 65 - 72.

Prakash, H. S. and Reddy, G. S. (1980). *Drosophila* fauna of Nagarhole, South India, including description of a new species (Diptera: Drosophilidae). *Proc. Indian Acad. Sci.* 89 (3), 235 - 241.

Rahman, S. M. Z. and Singh, V. K. (1969). A survey of Bhagalpur population of *Drosophila. Dros. Inf. Serv.* 44, 70.

Ranganath, H. A. and Krishnamurthy, N. B. (1972). Preliminary survey of *Drosophila* in Biligirirangana hills (Mysore state, India). *Dros. Inf. Serv.* 48, 132.

Reddy, G. S. and Krishnamurthy, N. B. (1968). *Drosophila rajasekari* – a new species from Mysore (India). *Proc. Indian Acad. Sci.* 68, 202 - 205.

Reddy, G. S. and Krishnamurthy, N. B. (1970). *Drosophila mysorensis* – a new species of *Drosophila* (Diptera: Drosophilidae) from Mysore, South India. *Journal biol. Sci.* 13 (2), 24 - 29.

Reddy, G. S. and Krishnamurthy, N. B. (1971). Preliminary survey of Drosophilids in Nilgiris and Kodaikanal ranges, *Dros. Inf. Serv.* 47, 116.

Reddy, G. S. and Krishnamurthy, N. B. (1973). Two new species of the *montium* subgroup of genus *Drosophila* (Diptera: Drosophilidae). *Orient. Insects,* 7 (2), 259 - 265.

Reddy, G. S. and Krishnamurthy, N. B. (1974). Systematics and distribution of *Drosophila* fauna of South India. *J. Mys. Univ.* 26(B), 54 - 64.

Sajjan, N. S. and Krishnamurthy, N. B. (1972). Report on two new species of *Drosophila* from Mysore. *Dros. Inf. Serv.* 45, 56 - 57.

Sajjan, N. S. and Krishnamurthy, N. B. (1973). *Drosophila neonasuta,* a new species of *Drosophila* from Mysore (Diptera: Drosophilidae). *Orient. Insects.* 7 (2), 267 - 270.

Sajjan, N. S. and Krishnamurthy, N. B. (1975). Two new Drosophilids from South India (Diptera: Drosophilidae). *Orient. Insects,* 9 (1), 117 - 119.

Sajjan, N. S. and Reddy, G. S. (1975). Two new species of *Drosophila* (Scaptodrosophila) (Diptera: Drosophilidae). *Orient. Insects,* 9 (1), 93 - 97.

Shivanna, N., Siddalingamurthy, G. S. and Ramesh, S. R. (1996). Larval pupation site preference and its relationship to the glue proteins in a few species of *Drosophila. Genome,* 39, 105 - 111.

Singh, A. (1970). Drosophilidae of South Andaman, India. *Dros. Inf. Serv.* 45, 106.

Singh, B. K. and Gupta, J. P. (1977). The subgenus *Drosophila* (*Scaptodrosophila*) in India (Diptera: Drosophilidae). *Orient. Insects,* 11 (2), 237 - 241.

Singh, B. K. and Bhatt, M. (1988). The Drosophilidae of Kumaon region with description of two new species and three new records. *Orient. Insects,* 22, 147 – 161.

Singh, B. K. and Negi, N. S. (1989). Drosophilidae of Garhwal region with description of one new species. *Proc. Zool. Soc. Cal.* 40, 19 – 25.

Singh, B. K. and Negi, N. S. (1992). Two new and one unrecorded species of Drosophilidae from Uttarakhand, India (Insecta: Diptera). *Senckenbergiana Biol.* 72, 321 – 327.

Singh, B. K. and Dash, S. (1993). Drosophilidae of Uttarakhand region with the description of one new species (Insecta: Diptera). *Proc. Zool. Soc. Cal.* 46 (2), 131 – 140.

Singh, B. K. and Negi, N. S. (1995). Further addition to the Drosophilidae fauna of Uttarakhand India (Diptera: Drosophilidae). *Entomol. Z.* 105 (21), 421 – 440.

Singh, B. K. and Dash, S. (1998). Drosophilidae of Kumaon region, India with the description of four new species (Insecta: Diptera). *Proc. Zool. Soc. Cal.* 51 (1), 45 – 56.

Singh, B. K., Dash, S. and Fartyal, R. S. (2000). Further additions to the genus *Leucophenga* from Kumaon, India (Insecta: Diptera). *Senckenbergiana Biol.* 80 (1/2), 149 – 154.

Singh, B. K. and Fartyal, R. S. (2002). Family Drosophilidae (Insecta: Diptera) in Kumaon region, India, with description of one new species and three new records. *Proc. Zool. Soc. Cal.*55 (1), 11 – 18.

Srinath, B. S and Shivanna, N. (2012). Note on A New Variant of *Drosophila daruma* Okada (Diptera: Drosophilidae) from Kalghatgi forest range, Dharwad, of Western Ghats. *Biosystematica,* 6(1), 39 – 42.

Srinath, B. S and Shivanna, N. (2014). *Drosophila* fauna of Dharwad District with a report of *Drosophila latifshahi* from South India. *Dros. Inf. Serv.* 97, 59 – 63.

Srinath, B. S. and Shivanna, N. (2017). First report of *D. Polychaeta* Patterson and Stone from India. *Dros. Inf. Serv.* 100, 58 – 62.

Srinath, B. S and Shivanna, N. (2019). First report of *Chymomyza vaidyaii* Okada (Diptera: Drosophilidae) from South India. *Dros. Inf. Serv.* 102, 43 – 46.

Sturtevant, A. H. (1919). A new species closely resembling *Drosophila melanogaster. Psyche,* 26, 153 – 155.

Sturtevant, A. H. (1921). The North American species of *Drosophila. Carnegie Inst. Wash. Publ.* 301, 1 – 150.

Sturtevant, A. H. (1939). On the subdivision of the genus *Drosophila. Proc. Natl. Acad. Sci.* 25, 137 – 141.

Sturtevant, A. H. (1942). The classification of the genus *Drosophila* with description of nine new species. *Univ. Tex. Publ.* 4213, 1 – 51.

Throckmorton, L. H. (1975). The phylogeny, ecology and geography of *Drosophila*. In: Handbook of Genetics (R. C. King ed.), Vol. 3. pp. 421–469. Plenum, NY.

Toda, M. J. and Peng, T. X. (1989). Eight species of the subgenus *Drosophila* (Diptera: Drosophilidae) from Guangdong Province, Southern China. *Zool. Sci.* 6 (1), 155 - 166.

Tsacas, L., Lachaise, D. and David, J. R. (1981). Composition and biogeography of the Afrotropical Drosophilid fauna. In: The Genetics and Biology of *Drosophila* (M. Ashburner, J. N. Thompson and H. L. Carson, eds.), Vol. 3a, pp. 197 - 259. Academic press, London.

Upadhay, K. and Singh, B. K. (2006). One new species and two new records of the subgenus *Drosophila* (Insecta: Diptera: Drosophilidae) from Kumaon Region, Uttaranchal, India. *Proc. Zool. Soc. Cal.*59 (2), 67 - 70.

Vaidya, V. G. and Godbole, N. N. (1971). Systematic study of Drosophilidae in Poona and neighbouring areas – I. *J. Univ. Poona*, 40, 49 - 61.

Vaidya, V. G. and Godbole, N. N. (1972). Systematic study of Drosophilidae in Poona and neighbouring areas – II. *J. Univ. Poona,* 42, 93 - 94.

Vaidya, V. G. and Godbole, N. N. (1973). Systematic study of Drosophilidae in Poona and neighbouring areas – III. *J. Univ. Poona,* 44, 41 - 43.

Vaidya, V. G. and Godbole, N. N. (1976). Systematic study of Drosophilidae in Poona and neighbouring areas – IV. *J. Univ. Poona*, 48, 85 - 92.

Val, F. C., Vilela, C. R. and Marques, M. D. (1981). Drosophilidae of the Neotropical region. In: The Genetics and Biology of *Drosophila* (M. Ashburner, J. N. Thompson and H. L. Carson, eds), Vol. 3a, pp. 123 - 168. Academic press, London.

Wheeler, M. R. (1981). Geographical survey of Drosophilidae: Nearctic species. In: The Genetics and Biology of *Drosophila* (M. Ashburner, J. N. Thompson and H. L. Carson, eds.), Vol. 3a, pp. 99 - 122. Academic press, London.

Wheeler, M. R. (1986). Additions to the catalog of the world's Drosophilidae. In: The Genetics and Biology of *Drosophila* (M. Ashburner, J. N. Thompson and H. L. Carson, eds.), Vol. 3e, pp. 395 - 409. Academic press, London.

Yassin, A. and David, J. R. (2010). Revision of Afrotropical species of *Zaprionus* (Diptera, Drosophilidae), with description of two new species and notes on internal reproductive structures and immature stages. *Zookeys*, 51, 33 - 72.

Chapter 7

Pteriodophytic Composition in Koppa Taluk, Central Western Ghats, South India

T.R. Parashurama[1,2], J. Deepa[2], K.G. Girish[2,3] and Prakash Kariyajjanavar[2,4]*

[1]Department of Botany,
Kumadvathi First Grade College, Shivamogga, Karnataka
[2]Panchavati Research Academy for Nature,
Kalamanji, Linganamakki, Sagara, Shivamogga, Karnataka
[3]Department of Environmental Science,
L.B and S.B.S. College, Sagara, Shivamogga, Karnataka
[4]Department of P.G Studies and Research in Environmental Science,
Gulbarga University, Kalaburagi, Karnataka
**Corresponding author e-mail: trparashurama@gmail.com*

Pteridophytes are vascular cryptograms reside of transitional position between bryophytes and spermatophytes and were the first plants to widely colonize the terrestrial ecosystem. In the present research work, forty six species of pteridophytes belonging to nineteen families were documented through distribution study in the forests of Koppa taluk, Chikmagalur district of Central Western Ghats. Adiantum philippense L was observed as higher importance value index followed by Thelypteris dentata (Forsk.) E.P.St.John., Pteris biaurita L, Adiantum raddianum C.Presl, Selaginella delicatula (Desv.ex Poir) Alston and Blechnum orientale L. The Shannon's diversity index value (H^1) and Simpson's diversity (D) values for pteridophytic species were indicated high diversity and species richness. The extensive deforestation and habitat *fragmentation in this region may pose serious threat to growth of pteridophytic communities. In this context, documentation and distribution studies of pteridophytic diversity are essential for conservation of threatened species.*

***Keywords**: Diversity indices, Conservation, Pteridophytes, Western Ghats.*

Introduction

India is diversified with rich pteridophytic flora due to its varied topographical climatic conditions and geographical position. About 9 per cent world Pteridophytic species are found in India. In India, many detailed pteridophytic survey have been reported by Beddome (1883, 1892) to Fraser Jenkins (2010) filled the large lacuna in our knowledge relating to the pteridology. Dixit (1984) has reported 1050 species and Chandra *et al.* (2008) put the number of pteridophytic species at 1150 from India. Later, Fraser Jenkins (2010) revised pteridophytic numbers to 1000 species in India including pteridophytes of Arunachal Pradesh (Fraser-Jenkins and Benniamin, 2010). Khullar (1994, 2000) listed 360 species of pteridphytes from Western Himalaya with 399 pteridophytes given by Fraser-Jenkins (2010). According to Manickam and Irudayaraj, (1992) Western Ghats supported 349 species out of 1100 to 1200 species of fern and fern allies in India. In Central Western Ghats, Karnataka region has richest pteridophytic diversity including Chikmagalur district (Sukumaran *et al.*, 2008, 2009; Deepa *et al.*, 2011, 2013a, 2013b; Nataraja *et al.*, 2011, Parashurama *et al.*, 2016, Fatima- Tu-Zahora-Jabeen and Nandashree B. M. 2019). The present study was under taken for fulfill the lacuna of information relating to the composition and distribution of different taxa of pteridophytes in forest region of Koppa taluk.

Study Area

The study area lies between 13° 32' 06 N latitude and between 75° 22' 05" E longitude. The altitude is 763 meters above m.s.l. and the annual rainfall exceeds 2500 mm. It encompasses area of 572.4 km^2 forest area.

Materials and Methods

A survey of pteridophytes in study area conducted during the period of 2015 to 2019. A total of 25 transects each measuring 50×2m were laid in forest of Koppa region. Different habitat forms of pteridophytes were recorded. In case of epiphytic form present on large tree considered as one colony and free floating hydrophytic form present in water sources also considered as one. Diagnostic features of all the specimens were studied and relevant field notes were made on fresh plant materials. Identification of specimens were made by referring to available literature and pteridophyte floras (Beddome, 1863, 1865, 1883; Clarke 1880, 1961; Blatter *et al.*, 1922; Tryon and Tryon, 1982; Bir, 1987; Khullar, 1994, 2000; Verma 2005 and 2008; Dixit, 1984; Chandra, 2008). Authentications of the species were done with the help of Indian Fern Society. All the collected specimens were properly processed and the herbariums have been deposited in the department of Applied Botany, Kuvempu University, Shankaraghatta. For nomenclature, Fraser-Jenkins (2008) has been followed.

Statistical Analysis

Data on various pteridophytic characters in different transect were collected and analyzed through statistical methods (Shukla, 2001; Tuomisto 2000; Deepa *et al.,* 2012). Density is an expression of the numerical strength of a pteridophytic species where the total number of individuals of each species in all transects are divided by the total number of transects studied. Frequency refers to the degree of dispersion of individual pteridophytic species in an area and usually expressed in terms of percentage occurrence. The samplings were made at different sampling units randomly in the study area and documented the species. Abundance is the study of the number of individuals of different species in the community per unit area. By transect method, samplings were made at random at several places and the number of individuals of each species was summed up for all the transects divided by the total number of transects in which the species occurred.

Relative density is the study of numerical strength of a species in relation to the total number of individuals of all the species. The degree of dispersion of individual species in an area in relation to the number of all the species occurred considered as Relative frequency. Importance Value Index is used to determine the overall importance of each species in the community structure. In calculating this index, the percentage values of the relative frequency, relative density and relative dominance are summed up together and this value is designated as the Importance Value Index or IVI of the species.

Based on the data of the occurrence of the species in the transects, Shannon's diversity index (H^1) was calculated which is represented by formula (H^1) = $-\Sigma p_i \ln p_i$, where $P_i = (n_i/N)$. Simpson's diversity index was calculated by formula (D) = $\Sigma [n_i (n_i-l)/N (N-l)]$, Where, '$n_i$' is the Number of individuals of the ith species and 'N' represented as total number of individuals.

Results and Discussion

In the present work, an attempt has made to study the distribution and species diversity of pteridophytes in Koppa taluk of Chikmagalur district. The phytosociological attributes such as abundance, density, frequency and their relative values (Table 7.1) and Importance Value Index were calculated in study area. The Shanon index is a measure of the average degree of "uncertainity", it is increases as the number of species increases and as the distribution of individuals among the species becomes even (John James, 1988). Simpson's index, which varies from 0 to 1, gives the probability that two individuals drawn at random from a population belong to the same species. If the probability is high that both individuals belongs to the same species, then the diversity of the community sample is low (John James, 1988).

Table 7.1: Checklist of Pteridophytes in Koppa Taluk, Chikmagalur District, Karnataka

Sl. No.	Name of the Species and Family	Den	Fre	Abu	Re. Den	Re. Ab	Re. Fre
1	*Adiantum concinnum* Humb. and Bonpl. *ex* Willd : Adiantaceae	4.16	0.36	11.56	3.52	3.12	2.93
2	*Adiantum latifolium* Lam. Adiantaceae	1.88	0.24	7.83	1.59	2.12	1.95
3	*Adiantum philippense* L.: Adiantaceae	13.68	0.88	15.55	11.58	4.20	7.17
4	*Adiantum raddianum* C.Presl : Adiantaceae	7.84	0.48	16.33	6.64	4.41	3.91
5	*Aleuritopteris anceps* (Blanf.) Panigrahi: Pteridaceae	2.12	0.28	7.57	1.79	2.04	2.28
6	*Angiopteris helferiana* C.Presl: Marattiaceae	1.92	0.24	8.00	1.63	2.16	1.95
7	*Arachniodes sledgei* Fraser-Jenk.: Dryopteridaceae	0.92	0.16	5.75	0.78	1.55	1.30
8	*Asplenium aethiopicum* (Burm.f.) Bech.: Aspleniaceae	0.48	0.12	4.00	0.41	1.08	0.98
9	*Asplenium formosum* Willd. : Aspleniaceae	1.52	0.2	7.60	1.29	2.05	1.63
10	*Asplenium phyllitidis* D.Don : Aspleniaceae	0.2	0.08	2.50	0.17	0.68	0.65
11	*Azolla pinnata* R.Br. subsp. *asiatica* R.M.K Saunders and K.Fowler : Azollaceae	0.16	0.08	2.00	0.14	0.54	0.65
12	*Blechnum orientale* L.: Blechnaceae	6.24	0.6	10.40	5.28	2.81	4.89
13	*Bolbitis subcrenatoides* Fraser-Jenk. : Lomariopsidaceae	0.2	0.04	5.00	0.17	1.35	0.33
14	*Cyathea gigantea* (Wall. *ex* Hook.) Holttum: Cyatheaceae	0.24	0.08	3.00	0.20	0.81	0.65
15	*Dicranopteris linearis* (Burm.f.) Underw.: Gleichenaceae	2.52	0.36	7.00	2.13	1.89	2.93
16	*Drynaria quercifolia* (L.) J.Sm. : Polypodiaceae	1.28	0.24	5.33	1.08	1.44	1.95
17	*Huperzia hamiltonii* (Spreng.) Trevis. : Lycopodiaceae	0.72	0.12	6.00	0.61	1.62	0.98
18	*Lepisorus nudus* (Hook.) Ching: Polypodiaceae	0.4	0.32	1.25	0.34	0.34	2.61
19	*Leptochilus lanceolatus* Fée : Polypodiaceae	0.12	0.04	3.00	0.10	0.81	0.33
20	*Lindsaea ensifolia* Sw.: Lindsaeaceae	1.36	0.24	5.67	1.15	1.53	1.95
21	*Lindsaea heterophylla* Dryand. :Lindsaeaceae	0.52	0.16	3.25	0.44	0.88	1.30
22	*Lycopodiella cernua* (L.) Pic.Serm.: Lycopodiaceae	0.32	0.12	2.67	0.27	0.72	0.98
23	*Lygodium flexuosum* (L.) Sw.: Lygodaceae	1.24	0.24	5.17	1.05	1.40	1.95

Sl. No.	Name of the Species and Family	Den	Fre	Abu	Re. Den	Re. Ab	Re. Fre
24	*Lygodium microphyllum* (Cav.) R.Br. :Lygodiaceae	0.2	0.08	2.50	0.17	0.68	0.65
25	*Marselia minuta* L. : Marsileaceae	0.24	0.04	6.00	0.20	1.62	0.33
26	**Microsorum zippelii* (Blume) Ching: Polypodiaceae	0.56	0.12	4.67	0.47	1.26	0.98
27	*Nephrolepis cordifolia* (L.) C.Presl: Oleandraceae	1.08	0.08	13.50	0.91	3.65	0.65
28	*Nephrolepis undulata* (Afzel. *ex* Sw.) J.Sm : Oleandraceae	0.72	0.12	6.00	0.61	1.62	0.98
29	*Odontosoria tenuifolia* (Lam.) J.Sm.: Lindsaeaceae	4.8	0.36	13.33	4.06	3.60	2.93
30	*Parahemionitis cordata* (Roxb. *ex* Hook. and Grev.) Fraser-Jenk.: Pteridaceae	0.72	0.08	9.00	0.61	2.43	0.65
31	*Pityrogramma calomelanos* (L.) Link.: Pteridaceae	1.92	0.36	5.33	1.63	1.44	2.93
32	*Poecilopteris virens* (Wall. Ex Hook. and Grev.) T.Moore : Lomariopsidaceae	4.56	0.72	6.33	3.86	1.71	5.86
33	*Pteridium revolutum* (Blume) Nakai: Dennstaedtiaceae	2.56	0.24	10.67	2.17	2.88	1.95
34	*Pteris argyrea* T.Moore : Pteridaceae	2.12	0.28	7.57	1.79	2.04	2.28
35	*Pteris biaurita* L.: Pteridaceae	9.48	0.92	10.30	8.03	2.78	7.49
36	*Pteris confusa* T.G.Walker : Pteridaceae	0.56	0.12	4.67	0.47	1.26	0.98
37	*Pteris pellucida* C.Presl: Pteridaceae	5.64	0.64	8.81	4.77	2.38	5.21
38	*Pteris vittata* L. subsp. *Vittata*: Pteridaceae	4.72	0.36	13.11	4.00	3.54	2.93
39	*Selaginella delicatula* (Desv.*ex*Poir) Alston : Selaginellaceae	4.32	0.12	36.00	3.66	9.72	0.98
40	*Selaginella tenera* (Hook. and Grev.) Spring : Selaginellaceae	5.48	0.24	22.83	4.64	6.17	1.95
41	*Selliguea montana* (Sledge) Hovenkamp : Polypodiaceae	0.2	0.12	1.67	0.17	0.45	0.98
42	*Tectaria polymorpha* Wall. *ex* Hook.: Tectariaceae	2.12	0.28	7.57	1.79	2.04	2.28
43	*Tecteria coadunata* (Wall. *ex* Hook. and Grev.) C.Chr. : Dryopteridaceae	2.76	0.32	8.63	2.34	2.33	2.61
44	*Thelypteris dentata* (Forsk.) E.P.St.John :Thelypteridaceae	11.44	0.72	15.89	9.69	4.29	5.86
45	*Thelypteris parasitica* (L) K. Iwats.	1.8	0.24	7.50	1.52	2.03	1.95
46	*Trichomanes companulatum* Roxb.: Hymenophyllaceae	0.08	0.04	2.00	0.07	0.54	0.33

Den: Density; Fre: Frequency; Abu: Abundance; Re. Den: Relative Density; Re. Fre: Relative Frequency; Re.Ab: Relative abundance; * Near threatened.

In Koppa taluk, 2,953 individuals were enumerated belonging to 19 families, 31 genera and 46 species. The pteridophytes of various habitats such as terrestrial, epiphytes, lithophytes and hydrophytes were found in this region.

Pteris is the largest genus having 5 species followed by Adiantum (4), Asplenium and Selaginella each of 3 species. There are17 out of 31 genera including single species in study area. *Adiantum philippense* L. is densely populated in study area followed by *Thelypteris dentata* (Forsk.) E.P.St.John, while, lowest for *Trichomanes companulatum* Roxb. *Selaginella delicatula* (Desv.*ex* Poir) Alston found to be more abundant followed by *Selaginella tenera* (Hook. and Grev.) Spring, *Adiantum raddianum* C.Presl *Adiantum philippense* L, *Lepisorus nudus* (Hook.) Ching recorded as of less abundance. The maximum frequency found in *Pteris biaurita* L while *Trichomanes companulatum* Roxb. observed as of low frequency.

Adiantum philippense L. was reported by 342 individuals showing higher importance value index of 22.95 and followed by *Thelypteris dentata* (Forsk.) E.P.St. John (19.84), *P. biaurita* (18.3), *Adiantum raddianum* C.Presl (14.96) and *Selaginella delicatula* (Desv.*ex* Poir) Alston (14.36). The Shannon's diversity index value (H^1) 3.25 and Simpson's diversity (D) = 0.052 values for pteridophytic species in Koppa taluk (Figure 7.1). Pteridaceae stands the dominant family of the study area with 12 species followed by Polyodiaceae with four species. *Pteris* and *Adiantum* were the largest genera with a maximum number of five and four species respectively.

Present study showing Pteridaceae was a dominant family followed by Polyodiaceae in contrast, the study made by Pinak *et al.* (2011) in Bark valley of Assam showed Polyodiaceae was largest family in followed by Thelypteridaceae, Aspidiaceae and Pteridaceae. Similar observation made by Sukumaran *et al.* (2009) also observed in Kanyakumari district of Tamilu Nadu. Present study indicated that heterogenous pteridophytic community, due to prediction of Raunkiaer (1934), the frequency class D is greater than A, B and C then the community is said to be heterogeneous. Mean rainfall positively correlated (p<0.01) with the species richness in the study area. Similarly, interpreting such bivariate relationships hampered by strong correlations between the climatic factors and ferns species richness (Jeremiah *et al.,* 1998). The extensive deforestation and habitat fragmentation activities would pose serious threat to growth of certain pteridophytic species. Henceforth documentation and distribution studies of pteridophytic diversity needs to be given top priority to help conservation resources and preservation of the threatened of species.

Anthropogenic activities are destructed and altered the habitat of different angiospermic species as well as pteridophytic species from the locality; besides this, it is also observed that encroachment of forest area for agriculture and developmental activities like expansion of roads and township extensions affecting these populations. In this context, present study provides fundamental information on a novel pteridophytic species composition and distribution in study area. This data could be used for designing conservation strategies in Koppa region of Central western Ghats.

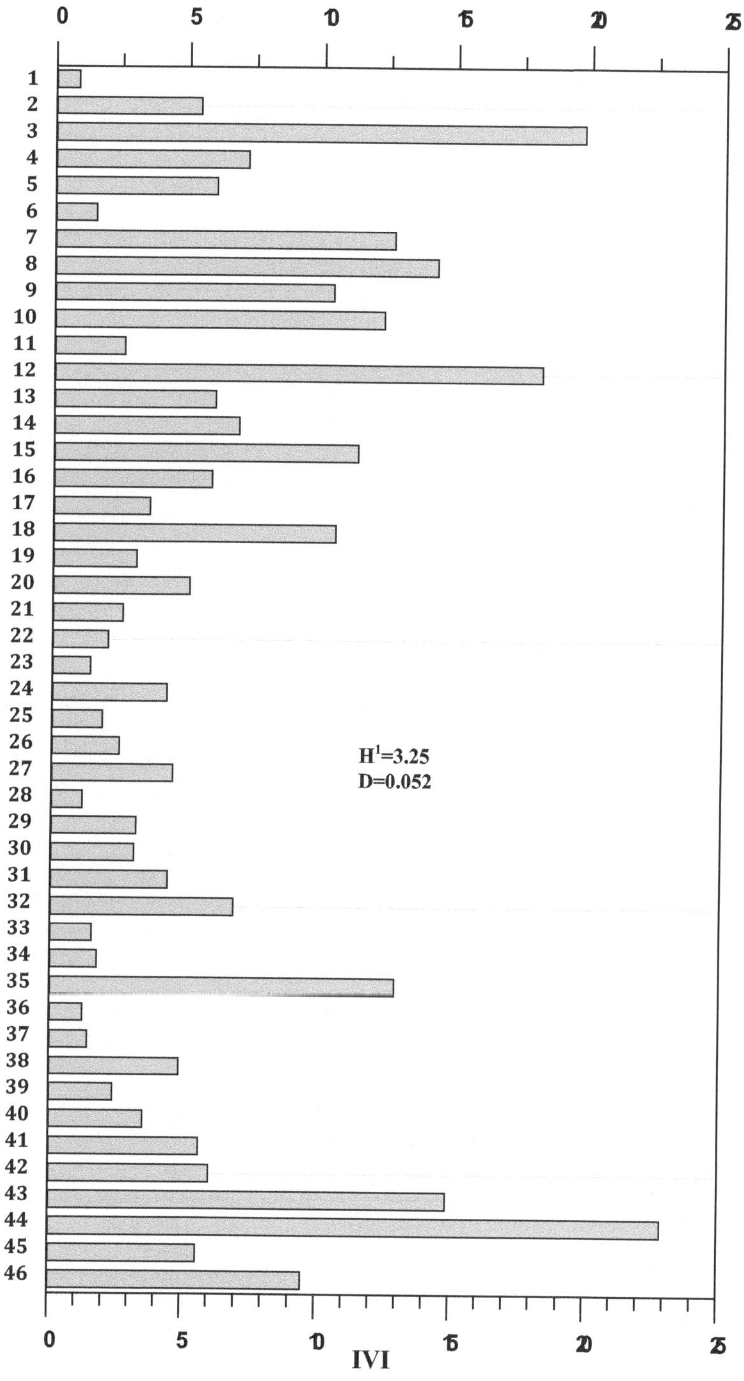

Figure 7.1: Diversity of Pteridophytes with Importance Value Index in Koppa Taluk of Chikmagalur District, Central Western Ghats, South India.

1	*Adiantum concinnum* Humb. and Bonpl. *ex* Willd
2	*Adiantum latifolium* Lam.
3	*Adiantum philippense* L.:
4	*Adiantum raddianum* C.Presl
5	*Aleuritopteris anceps* (Blanf.) Panigrahi
6	*Angiopteris helferiana* C.Presl
7	*Arachniodes sledgei* Fraser-Jenk.
8	*Asplenium aethiopicum* (Burm.f.) Bech.
9	*Asplenium formosum* Willd.
10	*Asplenium phyllitidis* D.Don
11	*Azolla pinnata* R.Br. subsp. *asiatica* R.M.K Saunders and K.Fowler
12	*Blechnum orientale* L.
13	*Bolbitis subcrenatoides* Fraser-Jenk.
14	*Cyathea gigantea* (Wall. *ex* Hook.) Holttum
15	*Dicranopteris linearis* (Burm.f.) Undrw.
16	*Drynaria quercifolia* (L.) J.Sm.
17	*Huperzia hamiltonii* (Spreng.) Trevis.
18	*Lepisorus nudus* (Hook.) Ching
19	*Leptochilus lanceolatus* Fée
20	*Lindsaea ensifolia* Sw.
21	*Lindsaea heterophylla* Dryand.
22	*Lycopodiella cernua* (L.) Pic.Serm.
23	*Lygodium flexuosum* (L.) Sw.
24	*Lygodium microphyllum* (Cav.) R.Br.
25	*Marselia minuta* L. : Marsileaceae
26	*Microsorum zippelii* (Blume) Ching
27	*Nephrolepis cordifolia* (L.) C.Presl
28	*Nephrolepis undulata* (Afzel. *ex* Sw.) J.Sm
29	*Odontosoria tenuifolia* (Lam.) J.Sm.
30	*Parahemionitis cordata* (Roxb. *ex* Hook. and Grev.) Fraser-Jenk.
31	*Pityrogramma calomelanos* (L.) Link.
32	*Poecilopteris virens* (Wall. Ex Hook. and Grev.)T.Moore
33	*Pteridium revolutum* (Blume) Nakai
34	*Pteris argyrea* T.Moore
35	*Pteris biaurita* L.
36	*Pteris confusa* T.G.Walker
37	*Pteris pellucida* C.Presl
38	*Pteris vittata* L. subsp. *Vittata*
39	*Selaginella delicatula* (Desv.*ex*Poir) Alston

40	*Selaginella tenera* (Hook. and Grev.) Spring
41	*Selliguea montana* (Sledge) Hovenkamp
42	*Tectaria polymorpha* Wall. *ex* Hook.
43	*Tecteria coadunate* (Wall. *ex* Hook. and Grev.) C.Chr.
44	*Thelypteris dentate* (Forsk.) E.P.St.John
45	*Thelypteris parasitica* (L) K. Iwats.
46	*Trichomanes companulatum* Roxb.

Note: H1: Shannon's diversity index value; D: Simpson's diversity; IVI: importance value index

Acknowledgments

The authors are grateful to Indian Fern Society and Department of Botany, Kuvempu University for help in identifications with valuable guidance and encouragement. Constant encouragement and motivation by Panchavati Research Academy for Nature (PRANA), Shivamogga, is greatly appreciated.

REFERENCES

Beddome, R. H. 1863-1865 *The Ferns of Southern India, tt.* 171. Gantz Bros., Madras

Beddome, R. H. 1865-1870 *The Ferns of British India, tt.* 345. Gantz Bros., Madras

Beddome, R. H. 1883 *Handbook to the Ferns of British India, Ceylon and the Malay Peninsula, pp.* 501. Thacker Spink and Co., Calcutta

Bir, S. S. 1987 Pteridophytic Flora of India: rare and endangered elements and their conservation, *Indian Fern J.* 4(1-2): 95-101

Blatter, E. and D'Almeida, J. E. 1922. *The Ferns of Bombay, pp.* 56-103. D.B. Taraporevala Sons and Co., Bombay

Chandra, S. 2000 *The Ferns of India* (Enumeration, Synonyms and Distribution), *pp.* 459. International Book Distributors, Dehra Dun

Chandra, S., Fraser-Jenkins, C. R., Kumari A and Srivastava A. 2008. A summary of the status of Threatened Pteridophytes of India *Taiwania* 53(2): 170-209

Clarke, C. B. 1880. A Review of the Ferns of Northern India. *Trans Linn Soc London* 2 Bot 1: 425-611

Deepa, J. Parashurama, T.R., Krishnappa, M. and Nataraja, S. 2011. Enumeration of Pteridophytes in Madhuguni Forest, Central Western Ghats, Karnataka, South India, Indian Fern J, 28: 112-119

Deepa J, T.R. Parashurama, M. Krishnappa and S. Nataraja 2013a. Pteridophytic Flora of Kemmangundi Forest, Karnataka, South India. Annals of Plant Sciences, pp 484 - 486.

Deepa, J., Parashurama, T.R., Krishnappa, M. and Nataraja, S. 2013b. Distribution of Pteridophytes in Kigga Forest, Central Western Ghats, Karnataka, South India. Indian Fern Journal. 30: 18-24.

Dixit R, D. 1984 *A Census of the Indian Pteridophytes*, Flora of India ser. IV, pp. 177. Botanical Survey of India, Howrah

Dixit, R.D. and Vohra, J.N. 1984. A Dictionaryof the Pteridophytes of India (Flora of IndiaSeries 4) Botanical Survey of India Publication, Department of Environment, Government of India, Botanical Garden, Howrah, pp. 1-177.

Fatima- Tu-Zahora-Jabeen and Nandashree B. M. 2019. Diversity of Pteridophytes and their medicinal uses in Mullayyangiri region of Chikkamagaliru taluk, Karnataka-State, India. International Journal of Scientific Research and Review. 8: 19-25

Fraser-Jenkins, C. R. and BENNIAMIN, A. 2010. ["2009"]. Fifty rarities and additions to the pteridophytic flora of Arunachal Pradesh, N.E. India, *Panjab Univ. Res. J., Sci.*, 59: 1-38

Fraser-Jenkins, C. R. 2008 Endemics and Pseudo-Endemics in Relation to the Distribution Patterns of Indian Pteridophytes Taiwania, 53(3): 264-292

Fraser-Jenkins, C. R. 2008. Taxonomic Revision of Three Hundred Indian Subcontinental Pteridophytes With a Revised Census-List, pp. 685. Bishen Singh Mahendra Pal Singh, Dehra Dun

Fraser-Jenkins, C. R. 2010. Nepal's little known pteridophytes, the hidden work of David Don, and the geography and distribution of Ind-Himalayan ferns with State lists, pp. 208. Website version: *www.groups.yahoo.com/group/Indian-Ferns*

Jeremiah S.L., Andrew B., and Rose B. 1998. Assessing fern diversity: relative species richness and its environmental correlates in Uganda, Biodiversity and Conservation, 7: 1387-1398.

John, A.L. and James, F.R. 1988. *Statistical Ecology: A Primer on Methods and Computing*. A Wiley-Interscience publication, p. 309.

Khullar, S. P. 1994 2000 An Illustrated Fern Flora of Western Himalaya. Vols.1 and 2. International Book Distributors, Dehra Dun

Manickam, V. S. and Irudayaraj, V. 1992. Pteridophytic Flora of the Western Ghats-South India, pp. 652. B I Publications Ltd. New Delhi

Nataraja, S., Deepa, J., Ramesh Babu, H. N. and Krishnappa, M. 2011. Pteridophytic survey in Agumbe forest of central Western Ghats, Karnataka Internat. J. Plant Sci., 6(2): 345-347

Parashurama T.R., Deepa J. and Prakash Kariyajjanavar, 2016. Pteridophytic Diversity in Mudigere Taluk, Central Western Ghats, Karnataka, South India. International Journal of Current Research, Volume 8, pp. 40339 -40342.

Pinak, D., Aniruuddha, S., Mrinal, K.B. and Mostafa, K.P. 2011. A review of the ferns of Bark Valley, Assam, India. *J. Taxon.Biodiv. Res.*, 5: 27-38.

Raunkiaer, C.1934. The Life forms of Plants and Statistical Plant Geography. Clarendon Press, Oxford.

Shukla, R. S and Chandel, P. S. 2001. Plant Ecology and soil Science, S.Chand and comp. Ltd. Ramnagar, New Delhi, pp: 110-141

Sukumaran, S. Jeeva, S. Raj, A. D. S. Kannan, D. 2008. Floristic Diversity, Conservation Status and Economic Value of Miniature Sacred Groves in Kanyakumari District, Tamil Nadu, Southern Peninsular India TurkJ Bot 32: 185-199

Sukumaran, S. Jeeva, S. Raj, A. D. S. Kannan, D. 2009. Diversity of Pteridopytes In Miniature Sacred Forests Of Kanyakumari District, Southern Western Ghats Indian Journal of Forestry, Vol.32(3): 285-290

Tryon, R. M. and Tryon, A. F. 1982. Ferns and Allied Plants with Special Reference to Tropical America Springer-Verlag New York

Tuomisto, H. and Poulsen, Axel Dalberg. 2000. Pteridopyte diversity and species composition in four Amazonian rain forest, Journal of Vegetation Science. Vol.11(3): 383-396

Verma, S. C. and Fraser-Jenkins, C. R. 2008. *Adiantum philippense* L., The correct name for *A. lunulatum* Burm.f., and its subspecie In 'Verma SC, Khullar SP and Cheema HK (Eds) *Perspectives in Pteridophytes*, Bishen Singh Mahendra Pal Singh, Dehra Dun, India, pp 65-92

Chapter 8

Occurrence of Malcofaunal Pest Complex in Agro Fields of Chikkamagluru Region, Karnataka

D.S. Somashekar[1], Shashikanth H. Majagi[2], M. Venkateshwarlu[3] and Dhananjaya S.G.[4]

[1]PG Studies in Zoology, I.D.S.G. Government College, Chikkamagluur – 577 102, Karnataka
[2]DOS in Zoology, VSK University, Ballari, Karnataka
[3]Department of PG Studies and Research in Zoology, Kuvempu University, Shankaraghatta – 577 451, Karnataka
[4]Govt. Science College, Chitradurga, Karnataka
e-mail: [1]drsomashekar7@gmail.com, [2]smajgi@rediffmail.com

Snails and slugs have a very important position in ecosystem as food source, pest and intermediate host and as bio indicators for environmental quality. The aim of the present study reveals the prevalence of snails and slugs in different agroecosystems of Chikkamagluru area. In the present study totally 8 species belonging to 6 genera were identified. Out of 8 species, 5 species were snails and remaining 3 species were slugs. In which A. fulia and D. leave found to be exotic species. According to the results analyzed during the present survey A.fulics was found at banana plantation in Mudigere taluk, and Areca in Chikkamagaluru and Belur taluk, and also found to occur in nurseries of Chikkamagaluru. During the present survey it was found that A. fulics, M. dussumari combination co-occur in Coffee, Banana and Areca plantation of study areas. While C. bistrialis, L. alte found to distributed in Maize and Ragi crop fields of Sukuruyuputnu region of Chikkamagaluru taluk. The brown slug M. dussumicn Gray, is found to occur in both Coffee arabica and robusta plantations. The percentage of occurrence of gastropods in Chikkamagaluru region is highest (57.27 per cent) compared to Mudigere (28.63 per cent) followed by Belur (14.09 per cent).

Keywords: *Gastropods, Chikkamagluru, A. fulics, Brown slug and Coffee plantation.*

Introduction

Land snails constitute about six per cent of the total species on Earth (Clark and May 2002). A large part of molluscan fauna in many tropical regions of the world is still poorly known. They form an important component of the forest ecosystem by recycling nutrients (Graveland *et al.*, 1994; Dunk *et al.*, 2004) and are the prey base for a number of small mammals, birds, reptiles, amphibians and other invertebrates, including carnivorous snails (Deepak *et al.*, 2010). In calcium poor habitats land snails can form an important source of calcium for other animals. Land snails also serve as an indicator of ecological conditions, and are very sensitive to climatic and ecological change (Simone 1999;). As early as 1839 Charles Darwin showed the value of land snails in studying environmental change (Naggs *et al.*, 2006). Terrestrial snails prove to be valuable research subjects for studies in evolutionary biology, biogeography, phylogeography, biodiversity, ecology and conservation biology (Schilthuizen *et al.*, 2007; Richards and Davison 2010). Highly diverse and narrowly distributed, land snails are good indicators of areas of conservation importance and endemicity when compared to widely distributed groups such as vertebrates (Moritz *et al.*, 2001).

The distribution and activity of land snails depends on several factors including precipitation, soil pH, soil Ca content, canopy density, *etc.* Calcium availability in the soil is a major limiting factor for their survival as it is required for their shell formation. Several studies have shown that Ca is positively correlated with species richness and density (Burch 1955; Aravind *et al.*, 2005). However, in the regions such as the Western Ghats, where the soil is usually acidic the snail richness is usually high but abundance is low.

Gastropods including slugs and snails are a very diverse group of molluscs. In India, 14 species of slugs have been reported to be pests of ornamental plants, vegetables and other crops causing damage by their rasping feeding (Kaur and Kaur, 2008a). Even though the damage caused by slugs to vegetable fields, fruit orchards, paddy, tea and mulberry plant is immense, insufficient work on their management exists in the literature as compared to that on insects, rodents and bird pests (Raut and Ghose, 1984).

Slugs are the major pests in agriculture, horticulture and floriculture, causing considerable damage to wheat, alfalfa, corn, soybean and tobacco (Donell *et al.*, 2008). Slugs injure plants by chewing holes in various sizes in the leaves and stems. These holes may be in the middle of the leaf or on the edge. The early seedlings stages are most susceptible to slugs; slugs can sometimes consume the entire seedling. Slugs are completely feed on small leaves while mature leaves show holes on them or eaten away around the edge. In karnataka the first incident of giant African snails was reported in coffee plantations at Bellarahalli village in Shanivarasanthe zone of Kodagu district of Karnataka, (Roobak Kumar *et al.*, 2018).

Gastropods have a very important position in our ecosystem as pest, intermediate host, food source and as bio indicators for environmental quality but the taxonomic work regarding snail species as not been carried out in this part of the world after 19th century (Javaria Altaf *et al.*, 2017). Madhyastha *et al.* (2004) reviewed the status of land snails of the Western Ghats and Mavinkuruve *et al.* (2004b) put a case forward for use of mollusc in conservation planning. Same authors made a provisional checklist of Karnataka state (Mavinkuruve *et al.*, 2004a). An attempt to address the diversity, endemism and geographical distribution patterns land snails of the Western Ghats has been made by Aravind *et al.* (2005) Apart from these studies, several papers appeared on the inventory of land snail diversity in several protected areas such as Rajiv Gandhi (Nagarahole) national park (Madyastha, 2004), Sharavathi wildlife sanctuary (Mavinkuruve *et al.*, 2005). The land snail fauna of the Western Ghats is better known when compared to other regions of India (Aravind *et al.*, 2005 and 2010).

Adequate study on the distribution and population of pestiferous malacofaunal (snails and slugs) lacking in Chikkamagaluru region. Hence, roving surveys were carried out to understand the distribution of these malacofauna particularly pestiferous molluscs in the environs of Chikkamagaluru region.

Materials and Methods

In order to investigate the level of snail infestation and to assess the damage, present survey was under taken in vicinity of Chikkamagaluru region.

A total of 440 snails were counted form 5 zones in and around Chikkamagaluru *viz.* Mudigere, Sakrepatna, Aldur and Belur during the period from July 2018 to October 2018, to know the diversity, intensity and infestation of snails. The snails were collected from different agro fields of different places mentioned above. The snails attached to vegetation were picked up from blunt forceps and were transfer to clean polythene bags partially closed with thread and transported to the Department of Zoology I.D.S.G. Government College, Chikkamagaluru for identification. Visual search method was employed to detect the pest snails. The specimen were collected and stored in 5 per cent formalin.

The identification of species was carried out by following the keys of Sundara Rajan and Latha (2014). Further farmers were interviewed to obtain feedback on the problem due to snail infestation to agriculture crops.

Study Area

Chikkamagaluru is situated roughly in the south western part of Karnataka state. A large are at this district in Malnadie forested hilly region of heavy rain fall.

Chikkamagaluru is located between 13° 15'53" north latitude and between 75° 04'46" and 76°21'50" east latitude. Its greatest length from cast to west in 138.4 km and from north to south east 88.5 km. It located at an elevation of 3400 feet. The temperature at ranges between 11-20°C during winter and 25-34° C during summer. The annual precipitation is 2000-2540 mm and considerably higher than the 750-1000mm.

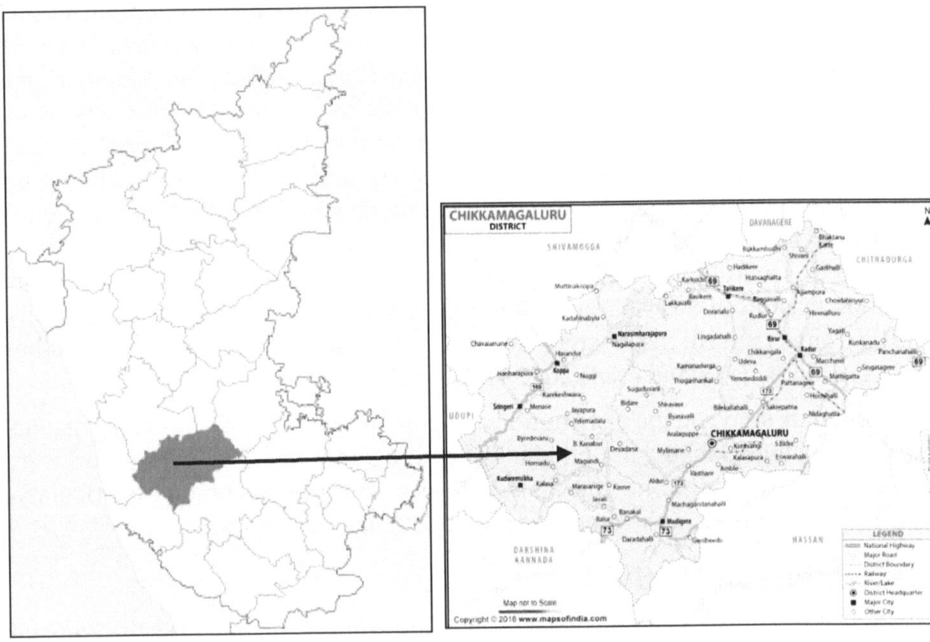

Figure 8.1: Map of Study Area.

Results and Discussion

Malcofauna have increasingly drawn attention of pest management in recent times due to their emergency as agricultural pests (Barker, 2002) snails and slugs as pest on ornamental and agricultural crops have been reported (Jayashankar *et al.*, 2015; Sridhar *et al.*, 2014).

In the present study totally 8 species belonging to 6 genera were identified. Out of 8 species, 5 species were snails and remaining 3 species were slugs. In which *A. fulia* and *D. leave* found to be exotic species (Table 8.1). *A. fulia* (Giant African Snail GAS) found to to infestation mainly on Areca plantation of Belur region *C. simensis* severely damaging the pepper plant and coffee plant in Mudigere and Chikkamagaluru region. The species *viz., D. leave, Macrochlamyus* species were found to infest on both *Coffee arabica* and *Coffee robusta*. *M. dussmieri* damaging both the cardamom and areca by eating its fresh seeds. *L. alte* and *Macrochlamyus species* feeding on ornamental plants.

Table 8.1: Check List of Malcofauna Pest Complex in Chikkamagaluru Region

Sl.No.	Species Name	Slug	Snail
1.	*Cryptozona siamensis*		X
2.	*Cryptozona bistrialis*		X
3.	*Achatina fulica* **		X
4.	*Mariallea dussumieri*	X	
5.	*Laevica ulisalte*	X	
6.	*Deroceras leave* **	X	
7.	Macrochlamys species		X
8.	Macrochlamys species		X

** : Exotic Species.

According to the results analyzed during the present survey *A. fulics* was found to infestation on banana crop in Mudigere taluk, and Areca in Chikkamagaluru and Belur taluk. The species is found to infest to the nursery ornamental plants in Chikkamagaluru (Tables 8.2 and 8.3). During the present survey it was found that *A. fulics M. dussumari* combination co-occur in Coffee, Banana and Areca plantation of study areas (Tables 8.1 and 8.2). While data on pest study of *C. Bistrialis, L. alte* found infestation on Maize and Ragi crops of Sakarayapatna region of Chikkamagaluru taluk. *D. laeve* was found in very low numbers in survey areas. It was found to infest to the ornamental plants of nursery.

Table 8.2: Infestation of Malcofauna on different Crops in Chikkamagaluru, Mudigere and Belur Region

Name of the crops		Infestation Rate of Malcofauna at Study Area		
		Chikkamagaluru Taluk	Mudigere Taluk	Belur Taluk
Areca	Not Observed	8	62	
Coffee	Arabica	147	Not Observed	Not Observed
	Robusta	Not Observed	99	Not Observed
Crops	Maize	15	Not Observed	Not Observed
	Ragi	3	Not Observed	Not Observed
	Banana	Not Observed	15	Not Observed
	Cardamom	Not Observed	4	Not Observed
Nursery		87	Not Observed	Not Observed
Total		**252**	**126**	**62**
Grand Total		**440**		

Table 8.3: Percentage of Occurrence of Malcofauna different Study Areas

Study Area	*No of Snails and Slugs*	*Percentage (per cent)*
Chikkamagaluru	252	57.27
Mudigere	126	28.63
Belur	62	14.09

The percentage of occurrence of Malcofauna pest in Chikkamagaluru region is highest (57.27 per cent) compared to Mudigere (28.63 per cent) followed by Belur (14.09 per cent) (Table 8.3 and Figure 8.2). The brown slug *M. dussumien* Gray, 1855 is found to feeding on the leaves of *Coffee arabica* and robusta. It is reported to be pestiferous on coffee in South India (Bhat and Shamanna, 1972), young rubber plants, green and wing beans in Sri Lanka (Naggs *et al.*, 2003), areca palms at Kasargod, Kerala (CPCRI, 2007), *Hibiscus rosa-sinensis, Anthuriumandraeanum, Piper nigrum, Theobroma cacao* and *Musa* sp. in Dakshina Kannada, Karnataka (Daniel and Vanavasan, 2009). Incidence of *M. dussumieri* inflicting damage to Marigold (*Tagetes* sp.) var in all the zones of Kodagu region (Tables 8.2 and 8.3).

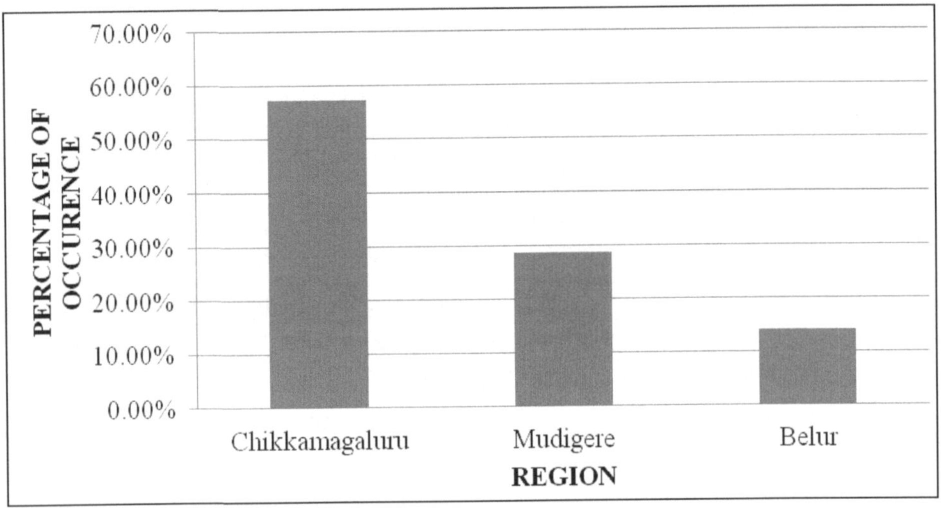

Figure 8.2: Percentage of Occurrence of Malcofauna different Study Areas.

The present observations indicate a need for more detailed survey of pestiferous terrestrial snails and emphasize the need to monitor invasive malacofauna in the region. Also, a thorough study to understand the impact of *A. fulica* on endemic snails needs to be undertaken. Further research quantifying the economic damage caused by malacofauna on different food crops in the study region has to be analyzed, as the impact of these snails is increasing necessitating suitable phytosanitary measures and Area Wide Pest Management (AWPM). Detecting pestiferous malacofaunal invasion early and initiating appropriate control measures would facilitate further spread and successful eradication.

These results therefore suggested that the extent of damage to agro horticulture crops depends on increased activity of slugs and snails in the field due to prevailing climatic conditions, stage and height of crop plant and shelter availability. In general, young seedling stage of plant was more vulnerable to slug attack. More damage was observed in the study period. So, it is suggested that slug control measures must be adopted during these months to minimize the slug and snails damage.

REFERENCES

Aravind N.A., K.P. Rajashekhar and N.A. Madhayystha. 2005. Species diversity, endemism and distribution of land snails of the Western Ghats, India. Records of the Western Australian Museum, Supplement 68: 31–38.

Aravind, N. A., Rajashekhar, K. P. and Madhyastha, N. A. 2010. A review of ecological studies on patterns and processes of distribution of land snails of the Western Ghats, India. Proceeding of World Congress of Malacology, p. 222.Barker, G. M. 2002. Molluscs as crop pests. CAB, International, Walling Forti Oxon 10 DE.UK. p. 468

Barker, G. M. 2002. Preface. In: Barker G. M (ed), Mollusc as Crop pests. CABI Publishing, Wallingford, 55-114.

Bhat, K. P and Shamanna, H.V. 1972. *Mariaella dussumieri* - a new pest of coffee in South India. J. Coffee Res., 2 (4): 28-29

Burch, J.B. (1955). 1962 How To Know the Eastern Land Snails. Wm. C. Brown, Dubuque, 214pp.

Clark, J.A. and R.M May (2002). Taxonomic bias in Conservation Research. Science 297: 191-192

CPCRI. 2007. Annual Report 2006-07, Central Plantation Crops Research Institute, Kasaragod - 671 124, Kerala, India, 124 p.

Daniel, M. and Vanavasan, N. K. 2009. The slug, *Mariaella dussumieri* Gray - an economically important pest of arecanut palms. Journal of Plantation Crops, 37(1): 88-90.

Deepak, V., K. Vasudevan and B. Pandav (2010). Preliminary observation on the diet of the cane turtle (*Vijayachelys silvatica*). Hamdaryad 34: 166–168.

Graveland, J.R, van der WAL, J.H. van Balen and A.J. van Noordwijk (1994). Poor reproduction in forest passerines from decline of snail abundance on acidified soils. Nature 368: 446–448.

Javaria Altaf, Naureen Aziz Qureshi, Muhammed Javed Iqbal Siddiqui. Taxonomic studies on the occurrence of the snails (Mollusca: Gastropods) in the agroecosystem. Journal of Biodiversity and Environmental Science (JBES). 2017; 10: 240-252.

Jayashankar M.M.S. Reddy and Ramakrishna 2015.Incidence of the common garden snail, *Macrochlamys indica benson*, 1832 (gastropoda: ariophantidae) in Bangalore region. The Bioscan 10(3): 1003-1006

Kaur, H. and Kaur, R. (2008a): Field evaluation of two additives for improving the acceptance of some molluscicide baits by the slug, Filicaulisalte Ferussac in vegetable fields at Malerkotla (Punjab). Environ and Ecol 26(4): 1602-06.

Madyastha, N.A., Mavinkurve, R. G. and Sandhya, P. S. 2004. Land Snails of Western Ghats. Envis. Wildlife Institute of India

Mavinkuruve. R. G., S. P. Shanbhag and N. A. Madhyastha. 2005. The land snails of Sharavati Wildlife Sanctuary. Records of Zoological Survey of India, 104: 123-131.

Mavinkuruve. R. G., S. P. Shanbhag, S.P. and N. A. Madhyastha. 2004a. Checklist of land snails of Karnataka. Zoos' Print Journal, 19: 1684-1686.

Mavinkuruve. R. G., S. P. Shanbhag, S.P. and N. A. Madhyastha. 2004b. Non-marine molluscs of Western Ghats. Zoos' Print Journal, 19: 1708-1711.

Moritz, C., K.S Richardson, S. Ferrier, G.B Monteith and J. Stanisic (2001). Biogeographical concordance and efficiency of taxon indicators for establishing conservation priority in a tropical rainforest biota. Proceedings of the Royal Society of London B. 268: 1875–1881.

Naggs, F., S. Panha and D. Raheem (2006). Developing Land Snail Expertise in South and Southeast Asia, a New Darwin Initiative Project. The Natural History Journal of Chulalongkorn University 6: 43–46

Naggs, F., Raheem, D.C., Mohan, P.B., Grimm, BRanwana, K.B. Kumburegama N B S 2003relicts and contemporary exotics: Faunal changes and survivorship in Srilanka's Snail fauna. Slugs and Snails: Agricultural, Veterinary and Environmental perspectives.2003 BCPC Symposium Proceedings No.80: 103-108

Raut, S. K. and Ghose, K. C. (1984): Pestiferous land snails of India. pp: 1-157. Technical Monograph No. 11. Zoological Survey of India, Calcutta.

Richards, P.M and A. Davison (2010).Adaptive Radiations: Competition Rules for Galápagos Gastropods. Current Biology 20: R28–R29.

Roobak Kumar A, Krishna Reddy P, Uma MS, Manjunath Reddy GV, Seetharama HG, Kurian Raphael P and M Dhanam2018.Occurrence of giant African snail, Achatinafulicabowdich in coffee growing areas of Karnataka and its management. Journal of Entomology and Zoology Studies; 6(4): 134-137

Schilthuizen, M., T-S.Liew, B.B. Elhahan and I. Lackman-Ancrenaz (2007). Effects of Karst Forest Degradation on Pulmonate and Prosobranch Land Snail Communities in Sabah, Malaysian Borneo. Conservation Biology 19: 949–954.

Simone, L.R.L. (1999). Mollusca Terrestres, p. 5: 3–8. In: Brandão, C.R. and E.M. Cancello, (Org.). Biodiversidade do Estado de São Paulo, Brasil: síntese do conhecimentoao final do século XX: InvertebradosTerrestres. FAPESP Editora, São Paulo, São Paulo, Brasil.

Soundararajan C, Latha BR. Veterinary Malacology. Department of Veterinary Parasitology, Madras Veterinary College, Chennai, Tamil Nadu. Prepared under ICAR - IDG Grant. 2014

Sridhar, V., Vinesh, L. S. and Jayashankar, M. 2014. Mapping the potential distribution of *Achatina fulica* (Bowdich) (Stylommatophora: Achatinidae) in India using CLIMEX, a bioclimatic software. Pest Management in Horticultural Ecosystems. 20(1): 14-21.

Chapter 9

Ichthyofauna Diversity in Relation to Water Quality of Lakes of Bangalore, Karnataka

*V. Sincy[1,4], R. Jaishanker[4], K.S. Asulabha[1,4] and T.V. Ramachandra[1,2,3]**

[1]Energy and Wetlands Research Group [CES TE15], Centre for Ecological Sciences,
[2]Centre for Sustainable Technologies (Astra)
[3]Centre for Infrastructure, Sustainable Transportation and Urban Planning (CiSTUP)
Indian Institute of Science, Bangalore – 560 012, Karnataka
[4]Indian Institute of Information Technology and Management-Kerala (IIITM-K), Thiruvananthapuram, Kerala
**Corresponding author e-mail: tvr@iisc.ac.in*

Ichthyofauna plays a vital role as consumers in the food chain and maintain the nutrient balance. The diversity and abundance of ichthyofauna depends on the nutrient levels in the aquatic ecosystem. The current study investigates ichthyofauna diversity and water quality across the lakes in Bangalore, Karnataka State, India. The order Cypriniformesis dominant in Bangalore lakes and 18 species were recorded in the current study, which is lower compared to the earlier report of 55 species in 1970. The decline in the native species and dominance of the exotic invasive species is mainly due to higher nutrient levels with the sustained inflow of untreated or partially treated wastewater to lakes. Lakes were earlier connected and now the loss of interconnectivity has also led to the disappearance of native species. Lakes of Bangalore are facing severe threat of pollution apart from the problems of encroachment of lakes, habitat loss, invasion of exotic species and over-harvesting of wetland resources. The quality of water was monitored and the overall quality is assessed through Water quality index (WQI). Principal component analysis was performed to understand the contribution of different physico-chemical parameters and their influence on ichthyofauna species diversity. The variations in the physico-chemical parameters have altered habitat conditions, affecting the diversity and abundance of ichthyofauna biodiversity. The study highlights the need to formulate strategies to sustain native species of ichthyofauna, which support the livelihood of local people through effective regulatory measures for preventing pollutants entering the water body.

Keywords: *Ichthyofauna, Lakes, Water pollution, Diversity, WQI.*

Introduction

Freshwater lakes play a significant role in the socioeconomic development of a region. The addition of chemicals, nutrients and organic contents to aquatic ecosystems through anthropogenic inputs creates an ecological imbalance affecting aquatic life. The sustained inflow of untreated wastewater from domestic and industrial areas to freshwater bodies leads to water pollution, eutrophication and heavy metal contamination (Ramachandra *et al.,* 2020a). Numerous studies have pointed out the major threats to wetlands that deteriorate water quality of freshwater resources, include water pollution, climate change, urbanization, land use and land cover changes, encroachments, agricultural practices, introduction of exotic species (Ramachandra and Aithal, 2016; Bassi *et al.,* 2014; Haidary *et al.,* 2013; Huang *et al.,* 2013; Bahar *et al.,* 2008; Prasad *et al.,* 2002). In addition to anthropogenic sources, the seasonal variations in natural processes like temperature, rainfall pattern and hydrological condition also influence water quality. The growth and development of fishes is dependent on both abiotic and biotic factors such as temperature, light, pH, oxygen level, feed, plankton population *etc.* (Bhatnagar and Devi, 2013). A change in the water quality parameters may affect the growth and survival of fish. Ichthyofaunaor fish is very sensitive to changes in water quality thus, considered as an effective biological indicator in aquatic ecosystems. Various climate change-induced threats on wetland fisheries include water stress, sedimentation, proliferation of aquatic weeds and loss of inter-connectivity among wetlands (Naskar *et al.,* 2018). Fishes are the major source of polyunsaturated fatty acids (PUFA)such aseicosapentaenoic acid (EPA) and docosahexaenoic acid (DHA) which are essential for human health (Barik, 2017; Sarma *et al.,* 2013). The fisheries sector provides livelihood for several families in India, in addition to fish constitute a valuable source of protein, vitaminsand minerals. Hence, assessment ofwater quality is crucial to understand the habitat quality of ichthyofauna.

Multivariate statistical technique, such as principal component analysis (PCA), help in the interpretation of complex data sets and provides better understanding of water quality of freshwater ecosystems (Zhang *et al.,* 2011; Wang *et al.,* 2017; Ramachandra *et al.,* 2018; Mamun *et al.,* 2021). PCA helps to analyze and interpret water quality data, understand temporal/spatial variations in water quality data and identify chief pollution sources of waterbodies, enabling water resource management (Barakat *et al.,* 2016; Islam *et al.,* 2017; Zeinalzadeh and Rezaei, 2017; Diamantini *et al.,* 2018). It provides information about the most meaningful parameters in the whole dataset, while removing data redundancy and explains the correlation between parameters with minimum loss of original information (Liu *et al.,* 2021; Mishra, 2010).

The water quality index (WQI) is widely accepted now to assess the water quality and determine the suitability of water for various purposes by integrating large set of water quality data, which includes physical, chemical and biological characteristics into a single number to express the quality of freshwater ecosystems (Ramachandra *et al.,* 2020b; Sharma and Kansal, 2011). WQI has been useful in water resource management worldwide for past few decades (Chabuk *et al.,* 2020; Bora and Goswami, 2017; Wu *et al.,* 2017; Jindal and Sharma, 2011). WQI is more effective than determining the trophic state index (TSI) or trophic level index (TLI) for aquatic ecosystems (El-Serehy *et al.,* 2018).

The studies on ichthyofauna diversity of lakes in Bangalore are scanty. Hence, the present study was designed to (i) record and map the ichthyofauna diversity; (ii) assess the water quality across the lakes in Bangalore, Karnataka; (iii) identify the pollution sources and pollution levels using multivariate statistical analysis (principal component analysis, PCA)and (iv) compute water quality index (WQI).

Materials and Methods

Description of the Study Area

Greater Bangalore (city of lakes) lies between the latitude 12°49’5”N to 13°08’32”N and longitude 77°27’29”E to 77°47’2”E, covering an area of 741 square kilometers. Bangalore landscape with undulating terrain forms three major watersheds namely Koramangala and Challaghatta Valley (K and C Valley), Vrishabhavathi Valley (V Valley) and Hebbal Valley (H Valley). Koramangala and Challaghatta Valley has the largest catchment area (255 square kilometers), followed by Hebbal Valley (207 square kilometers) and Vrishabhavathi Valley (165 square kilometers). The lakes in Bangalore are interconnected and provide goods and services to sustain livelihood of people. The interconnected lakes in K and C and H valley join Dakshina Pinakini River whereas lake systems in V valley join Arkavathi River, a tributary of Cauvery River (Ramachandra *et al.,* 2017). The current study involves field survey, water quality analysis and recording ichthyofauna diversity in 36 lakes of Greater Bangalore (Figure 9.1).

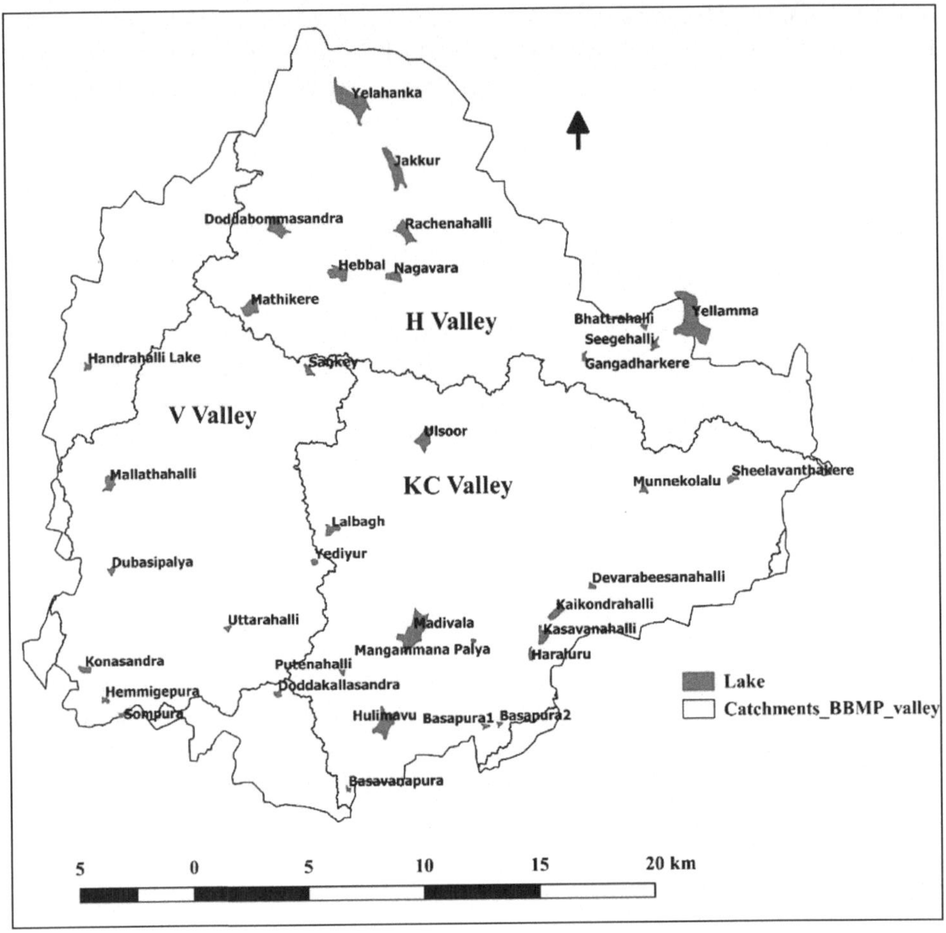

Figure 9.1: Study Area - Lakes of Greater Bangalore, Karnataka, India.

Ichthyofauna Sample Collection

Fish specimen were collected from the select lakes of Bangalore with the help of gill net or cast net by local fishermen during the study period 2018 - 2019. The photographs of ichthyofauna were taken prior to the specimen preservation. The specimens were preserved in 10 per cent formaldehyde solution and transported to the laboratory for taxonomic analysis. Collected fish specimen were identified up to the species level by using standard taxonomic keys, reference books (Jayaram, 1999; Talwar and Jhingran, 1991) and virtual database - Fish Base (http://fishbase.org). The species distribution map was generated with the help of QGIS open-source GIS software.

Water Sample Collection and Laboratory Analyses

Water samples were collected from the 36 lakes for physico-chemical analysis in the clean and sterile polypropylene bottles during the year 2018 - 2019. The monitored parameters included water temperature (WT), pH, dissolved oxygen (DO), total dissolved solids (TDS), electrical conductivity (EC), chloride (Cl), total hardness (TH), calcium (Ca), magnesium (Mg), total alkalinity (TA), chemical oxygen demand (COD), biochemical oxygen demand (BOD), nitrate (N) and ortho-phosphate (OP). The parameters such as water temperature, pH, dissolved oxygen, total dissolved solids and electrical conductivity were measured on site using portable meter (Extech) while rest of physico-chemical parameters were estimated in laboratory. The lake water samples were analyzed using standard methods according to APHA (2005) and every experiment were carried out in triplicates. In this study, QGIS open-source GIS software was used for generating maps of various physico-chemical parameters of investigated 36 lakes in Bangalore.

Estimation of WQI of Lakes

During the present study, weighted arithmetic water quality index (WQI) was computed for the 36 lakes in Bangalore considering ten physico-chemical parameters such as pH, total dissolved solids, electrical conductivity, dissolved oxygen, total hardness, calcium, magnesium, chloride, total alkalinity and nitrate (Devi *et al.*, 2015; Sincy *et al.*, 2016).

Statistical Analysis

In the present study, multivariate analysis was carried out using PAST3 software. The principal component analysis (PCA) of the water quality data was performed to better understand water quality of the monitored lakes. Experimental data was normalized using z-scale transformation for the principal component analysis, which reduces the influence of variance of different parameters, eliminate the influence of dissimilar units of measurement of the analyzed parameters (Singh *et al.*, 2004), renders the data dimensionless (Singh *et al.*, 2020) and helps to assess the association among monitored variables (Vega *et al.*, 1998).

Results and Discussion

Ichthyofauna Diversity of Lakes

The present study records 18 species of ichthyofauna belonging to 4 orders, 7 families and 14 genera (Figure 9.2). The ichthyofauna belonging to order (Figure 9.3a) Cypriniformes were dominant in lakes (41 per cent) followed by Perciformes (33 per cent), Cyprinodontiformes (15 per cent) and Siluriformes (11 per cent).

The order Cypriniformes consists of species namely *Catla catla, Labeo rohita, Ctenopharyngodon idella, Cyprinus carpio, Cirrhinus mrigala, Labeo fimbriatus, Puntius ticto* and *Hypophthalmichthys molitrix*. Cyprinodontiformeswas represented by species such as *Gambusia affinis* and *Poecilia reticulata*. Perciformes was represented by *Oreochromis mossambicus, Oreochromis niloticus, Channa punctata, Channa striata* and *Parambassis ranga*. Siluriformeswas represented by *Clarias gariepinus, Clarias batrachus* and *Heteropneustes fossilis*. Similar results were observed in other lakes, for example, the order Cypriniformes were dominant in Mallasandralake of Tumakuru (Nayaka, 2018); Kelagerilake of Dharwad (Kamble and Ganesh, 2016) and Nagaram tank (Ramulu and Benarjee, 2013).

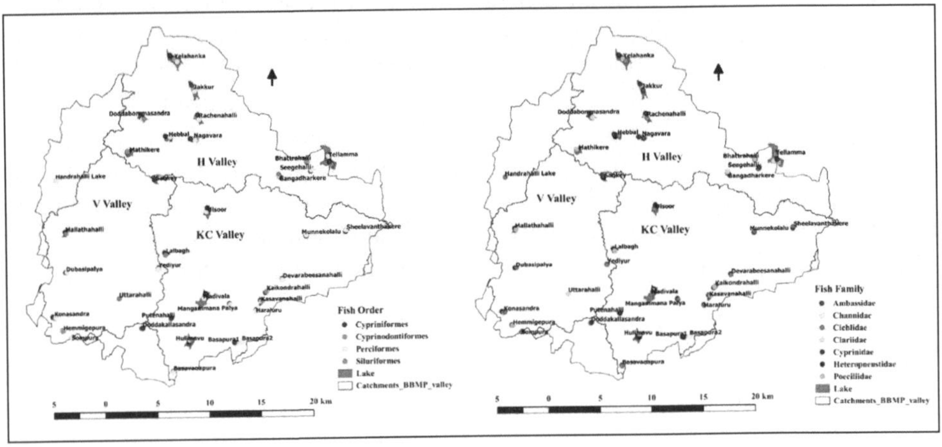

Figure 9.2: Distribution of Ichthyofauna (Order wise and family wise) in Lakes of Bangalore.

Fish species belonged to 7 families (Figure 9.3b) namely, Ambassidae, Channidae, Cichlidae, Clariidae, Cyprinidae, Heteropneustidae and Poeciliidae. Among these, Cyprinidae was dominated in 41 per cent of monitored lakes, followed by Cichlidae (28 per cent), Poeciliidae (15 per cent), Clariidae (9 per cent), Channidae (3 per cent), Ambassidae (1 per cent) and Heteropneustidae (1 per cent).

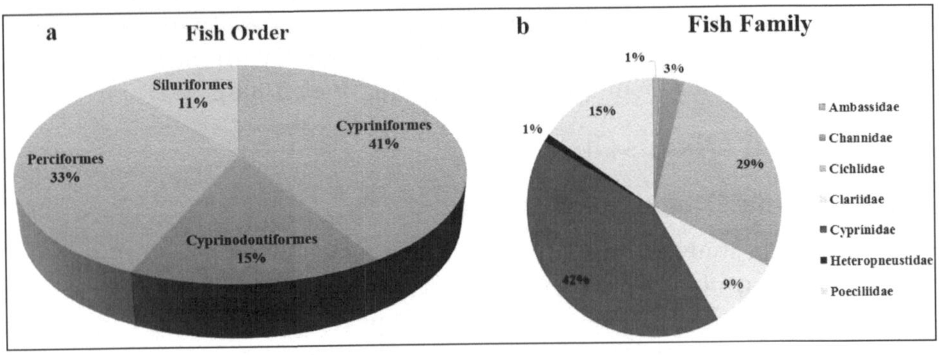

Figure 9.3: The Percentage Occurrence of Fish Orders (a) and Families (b) in Lakes of Bangalore.

Cyprinus carpio, Ctenopharyngodon idella, Gambusia affinis, Oreochromis sp., *Clariasgarie pinus* and *Hypophthalmichthys molitrix* are exotic fishes recorded in the current study. *Clarias gariepinus* is a hardy species and can tolerate both oxygenated as well as poorly oxygenated waterbodies (Ogundiran *et al.*, 2010).

Table 9.1: Ichthyofauna Diversity in Lakes of Bangalore, Karnataka

Sl.No.	Fish Species	Common Name	Order	Family	Genus
1	*Catla catla*	Catla	Cypriniformes	Cyprinidae	Catla
2	*Channa punctata*	Spotted Snakehead	Perciformes	Channidae	Channa
3	*Channa striata*	Striped Snakehead	Perciformes	Channidae	Channa
4	*Cirrhinus mrigala*	Mrigal	Cypriniformes	Cyprinidae	Cirrhinus
5	*Clarias batrachus*	Walking catfish/Magur	Siluriformes	Clariidae	Clarias
6	*Clarias gariepinus*	African catfish	Siluriformes	Clariidae	Clarias
7	*Ctenopharyngodon idella*	Grass carp	Cypriniformes	Cyprinidae	Ctenopharyngodon
8	*Cyprinus carpio*	Common carp	Cypriniformes	Cyprinidae	Cyprinus
9	*Gambusia affinis*	Mosquitofish	Cyprinodontiformes	Poeciliidae	Gambusia
10	*Heteropneustes fossilis*	Stinging catfish	Siluriformes	Heteropneustidae	Heteropneustes
11	*Hypophthalmichthys molitrix*	Silver carp	Cypriniformes	Cyprinidae	Hypophthalmichthys
12	*Labeo fimbriatus*	Fringed-lipped peninsula carp	Cypriniformes	Cyprinidae	Labeo
13	*Labeo rohita*	Rohu	Cypriniformes	Cyprinidae	Labeo
14	*Oreochromis mossambicus*	Mozambique tilapia	Perciformes	Cichlidae	Oreochromis
15	*Oreochromis niloticus*	Nile tilapia	Perciformes	Cichlidae	Oreochromis
16	*Parambassis ranga*	Indian glassy fish	Perciformes	Ambassidae	Parambassis
17	*Poecilia reticulata*	Guppy	Cyprinodontiformes	Poeciliidae	Poecilia
18	*Puntius ticto*	Ticto Barb	Cypriniformes	Cyprinidae	Puntius

Gambusia affinis and *Poecilia reticulata* feed on larvae and is used for mosquito control (Mahapatra *et al.*, 2014). The invasion of exotic species results in loss of native and endemic species (Singh *et al.*, 2013; Sandilyan, 2016). Exotic fish species were introduced in India for aquaculture, aquarium, sport fishing, weed control, *etc.* and the introduction of exotic species led to the decline in native species of fish and biodiversity (Kumar, 2000).

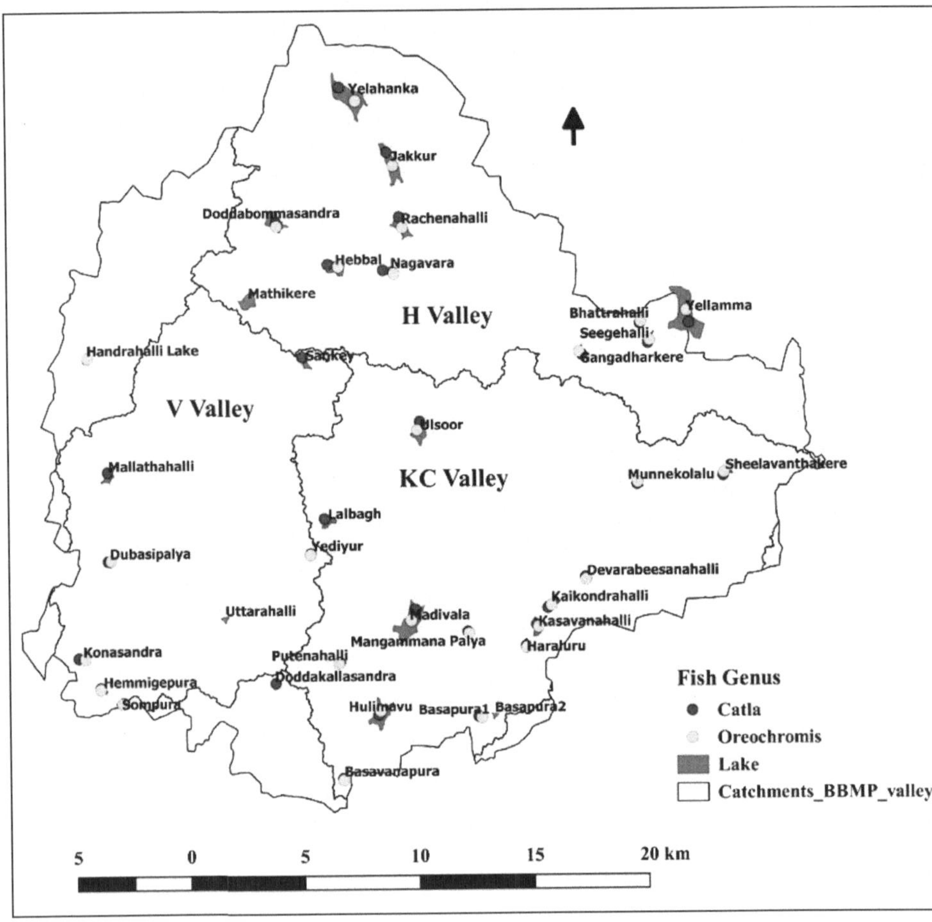

Figure 9.4: Distribution of Catla and Oreochromis in Bangalore Lakes.

Among the 14 genera, Catla was found in 32 lakes and Oreochromis was recorded in 29 lakes in the current investigation (Figures 9.4 and 9.5). Based on the diet (Table 9.2), fishes are grouped as herbivorous, omnivorous, larvivorous, planktivorous and carnivorous. For example, *Labeo rohita* (column and bottom feeder) prefers plant material and decaying vegetation. *Catla catla* (surface feeder) feed mainly on phytoplankton whereas *Cirrhinus mrigala* (bottom-feeder) feed mainly on decaying vegetation (Chattopadhyay, 2017). *Oreochromis niloticus* are opportunistic feeders and feed on detritus, phytoplankton, crustacean, aquatic insects, small fish, zooplankton and macrophytes (Mohamed and Al-Wan, 2020; Chatterjee *et al.*, 2015).Under conservation status as per IUCN (2010),one species was categorized as critically endangered, one species under near threatened, 13 under least concern and one species under vulnerable category (Table 9.2).

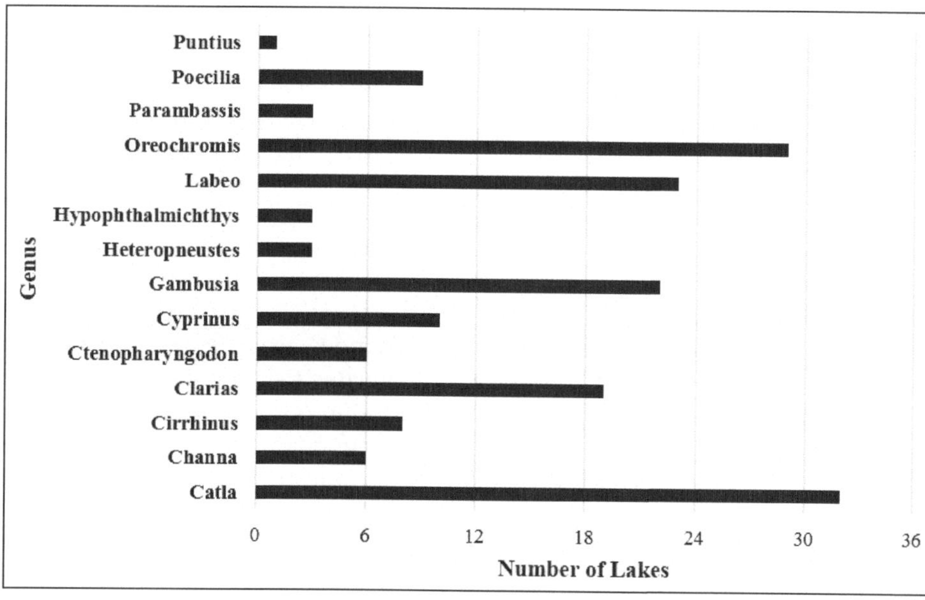

Figure 9.5: Fish Genera found in Bangalore Lakes

Table 9.2: List of Ichthyofauna, their IUCN Status, Diet Pattern and their Benefits

Sl.No.	*Fish Species*	*IUCN**	*Diet*	*Benefits*
1	*Catla catla*	LC	Omnivorous	Food fish
2	*Channa punctata*	LC	Carnivorous	Food fish
3	*Channa striata*	LC	Carnivorous	Food fish
4	*Cirrhinus mrigala*	LC	Omnivorous	Food fish
5	*Clarias batrachus*	LC	Omnivorous	Food fish
6	*Clarias gariepinus*	LC	Omnivorous	Food fish
7	*Ctenopharyngodon idella*	NE	Herbivorous	Control of weeds/ Food fish
8	*Cyprinus carpio*	CR	Omnivorous	Food fish
9	*Gambusia affinis*	LC	Larvivorous	Mosquito-control/ Ornamental
10	*Heteropneustes fossilis*	LC	Omnivorous	Food fish
11	*Hypophthalmichthys molitrix*	NT	Planktivorous	Food fish/controls pest population

Sl.No.	Fish Species	IUCN*	Diet	Benefits
12	*Labeo fimbriatus*	LC	Herbivorous	Food fish
13	*Labeo rohita*	LC	Omnivore	Food fish
14	*Oreochromis mossambicus*	VU	Omnivorous	Food fish
15	*Oreochromis niloticus*	LC	Omnivorous	Food fish
16	*Parambassis ranga*	LC	Carnivorous/ Larvivorous	Ornamental fish
17	*Poecilia reticulata*	NE	Omnivorous	Ornamental fish
18	*Puntius ticto*	LC	Planktivorous	Food fish/ Ornamental fish

*LC: Least Concern; NE: Not Evaluated; CR: Critically Endangered; NT: Near Threatened; VU: Vulnerable.

Indian major carps such as *Catla catla, Cirrhinus mrigala* and *Labeo rohita* contribute 70 to 75 per cent of the total freshwater fish production and have high economic importance (Jayasankar, 2018). Genus such as Clarias, Channa and Heteropneustes are air breathing fishes with good market value for live fish (Thirumala *et al.,* 2011).Dishes prepared with fish flour (of *Oreochromis niloticus*) are rich in proteins, lipids, essential amino acids, ash and polyunsaturated fatty acids (PUFAs) (Alam and Aslam, 2020), which highlights the need to conserve fish fauna to sustain food and medicine to the dependent population.

Water Quality of Lakes

Health of aquatic ecosystems are assessed based on water quality considering the physical, chemical and biological characteristics of water. Evaluation of these parameters will aid in understanding the suitability of lake water for biotic consumption and health of a particular aquatic ecosystem. Fish is very sensitive to variations in water quality and hence considered as biological indicator in aquatic ecosystems. Thus, it is necessary to understand the water quality of lakes or habitat conditions of fishes. Physical, chemical and biological parameters assessed in sampled lakes are discussed next.

Water Temperature and pH

Water temperature influences both the physico-chemical variables and biological activities in the aquatic ecosystems. Temperature affects the microbial (algae and bacteria) activities in water as it influences the organic matter decomposition and nutrient cycling. Temperature also affects the solubility of gases like O_2, CO_2 and NH_3 (Siriwardana *et al.,* 2019). Water temperature in the current study varied between 23.5°C to 34.2°C (Figure 9.6). pH varied among the monitored lakes and found to range between 7.2 to 10.19, indicated an alkaline nature of lake water (Figure 9.6).

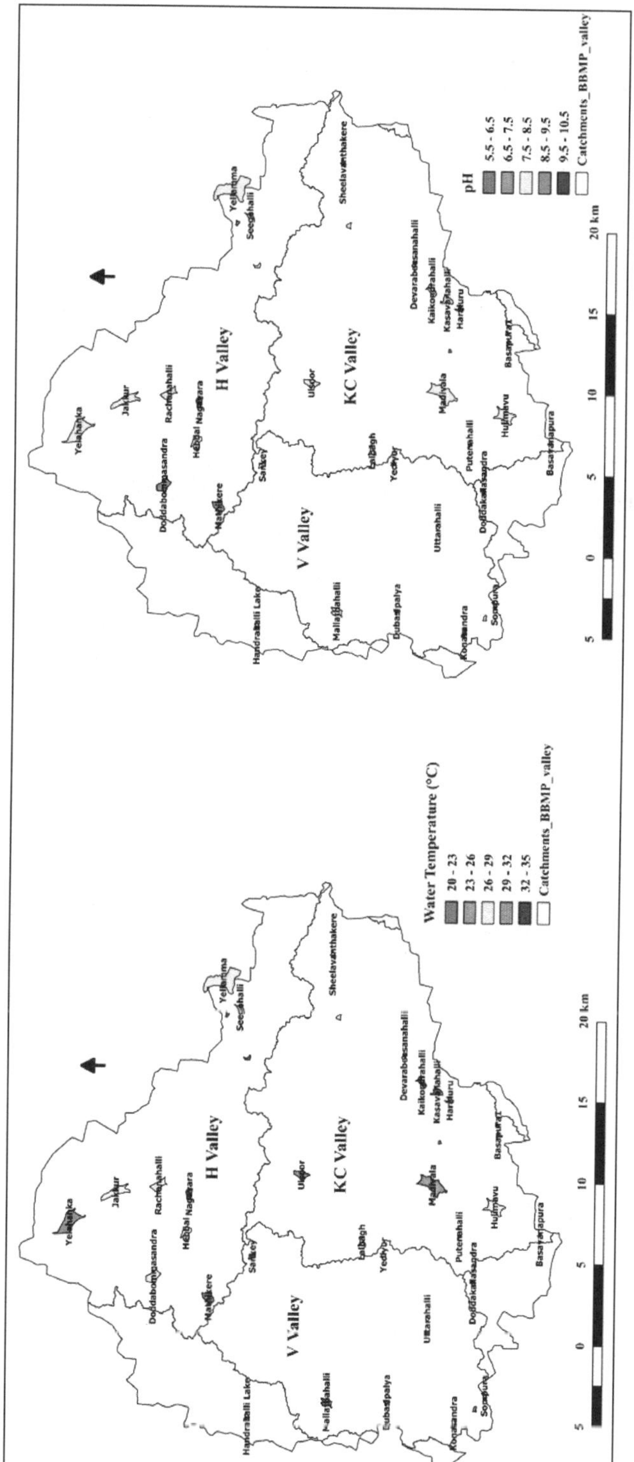

Figure 9.6: The Variation of Water Temperature and pH across Lakes of Bangalore.

pH regulates many biological processes and biochemical reactions. In case of pH, the desirable range for fish culture is 6.5 - 9.0. Doddakallasandra had high levels of pH among all the lakes due to high photosynthetic activities by phytoplankton. The photosynthesis raises the pH whereas respiration and decomposition processes lowers pH in lake water. Fish have an average blood pH of 7.4, so the acceptable range for fish culture would be 6.5 - 9.0. The pH in water ranging from 4.0 - 6.5 and 9.0 - 11.0 is stressful for fish growth and reproduction. Fish death can occur at pH level of <4 or>11 (Ekubo and Abowei, 2011).

Total Dissolved Solids (TDS) and Electrical Conductivity (EC)

Total dissolved solids in natural waters are mainly contributed by the presence of carbonates, bicarbonates, sulphates, chlorides, phosphates and nitrates of magnesium, calcium, sodium, potassium, iron *etc.* and tiny amounts of organic matter (Ramachandra *et al.,* 2014). Electrical conductivity gives the measure of the capacity of water to conduct an electric current. When the concentration of dissolved ions in water increases then, electrical conductivity also increases. The value of total dissolved solids and electrical conductivity in lakes ranged from 152.5 mg/L to 1548 mg/L and 306.67μS/cm to 2814 μS/cm, respectively (Figure 9.7). Lakes such as Basapura-1 and Basapura-2 had higher ionic contents with higher TDS and EC levels. The increased levels of dissolved and suspended solids will decrease the dissolved oxygen levels in aquatic ecosystems (Yýlmaz *et al.,* 2020). The optimum conductivity level for fish culture differs among different species as they vary in their capability to maintain osmotic pressure (Kumar *et al.,* 2017).

Turbidity and Dissolved Oxygen (DO)

Turbidity occurs due to the presence of various suspended and colloidal matter in lake water such as clay, silt, inorganic matter, organic matter, plankton and tiny micro-organisms (Ramachandra *et al.,* 2018). Turbidity in monitored lakes was found to range between10.14 NTU - 306.33 NTU (Figure 9.8) .High turbidity in lakes result in low light penetration to bottom levels, reduces photosynthetic rate and may even cause fish death due to clogging of gills (Saraswathy *et al.,* 2015). The concentration of dissolved oxygen in lake water varies with temperature, amount of organic waste, sediment quality/quantity, turbulence, photosynthetic rate and respiration/decomposition (Devi *et al.,* 2015). The solubility of dissolved oxygen decreases under low atmospheric pressure and high saline conditions (Rouse, 1979). The dissolved oxygen levels ranged between 1.52 mg/L - 11.27 mg/L in lakes with minimum value in Gangadharkere while maximum value in Jakkur lake (Figure 9.8).

Biochemical Oxygen Demand (BOD) and Chemical Oxygen Demand (COD)

Biochemical oxygen demand is the amount of oxygen utilized by micro-organisms in stabilizing the organic matter (Davraz *et al.,* 2019). The chemical oxygen demand is the amount of oxygen equivalent of the organic matter in lake

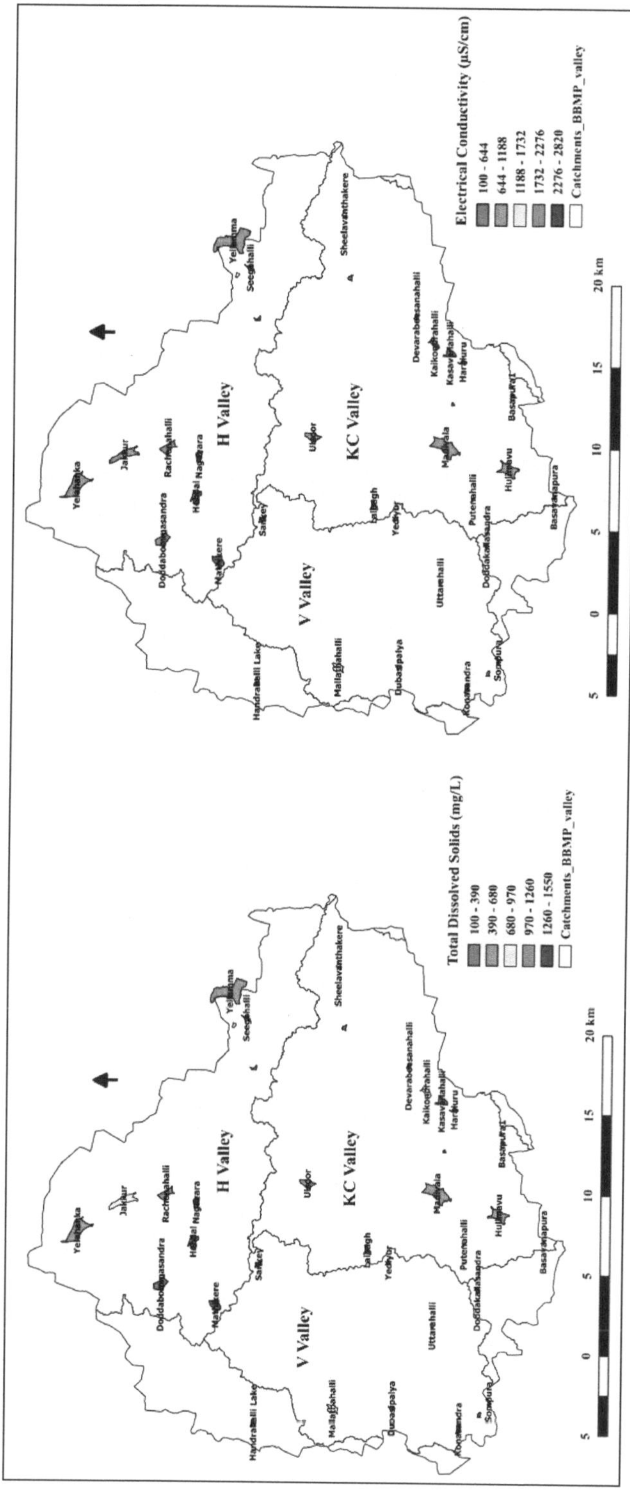

Figure 9.7: Variation of Total Dissolved Solids and Electrical Conductivity in Lakes of Bangalore.

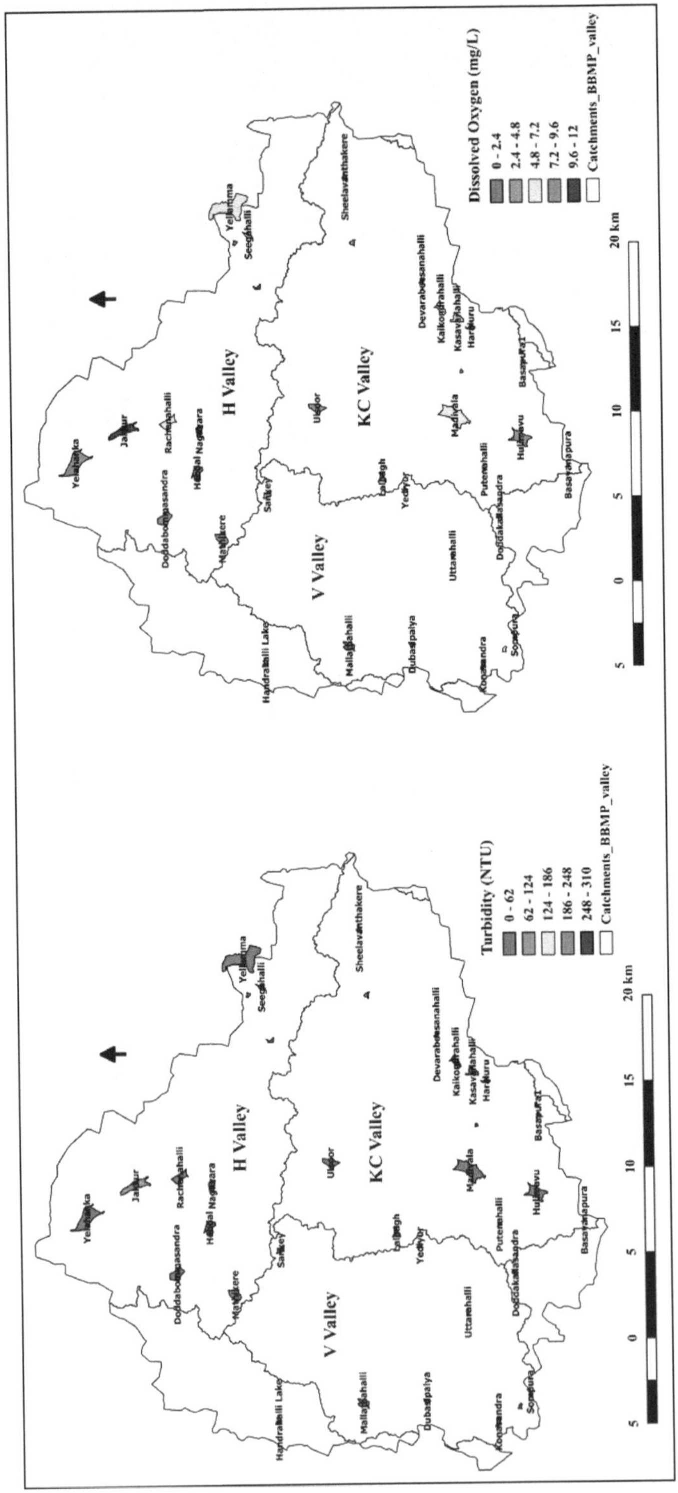

Figure 9.8: The Turbidity and Dissolved Oxygen Levels in Lakes of Bangalore.

water that is susceptible to oxidation with the help of a strong chemical oxidant. The BOD and COD of lake water samples ranged between 7.45 mg/L - 99.59 mg/L and 11.33 mg/L - 150 mg/L, respectively (Figure 9.9). Both BOD and COD of lake water increases with an increase in the organic matter content due to organic pollution from point as well as non-point sources.

Total Alkalinity and Chloride

Total alkalinity provides the measure of the ability of water to neutralize acids. Alkalinity is due to the presence of carbonates, bicarbonates and hydroxides of calcium, magnesium, potassium, sodium and salts of borates, silicates, phosphates, *etc.* (Sarkar *et al.,* 2020, Qureshimatva *et al.,* 2015). The salts of sodium (NaCl), potassium (KCl) and calcium ($CaCl_2$) contribute chlorides in lake water. The total alkalinity and chloride in sampled lakes were recorded as 81.33 mg/L - 684.33mg/L and 25.56 mg/L - 1165.82 mg/L, respectively (Figure 9.10). The desired level of alkalinity for fish culture is between 50-150 mg/L whereas the desirable level of chlorides for catfish production is > 60 mg/L (Rajbongshi *et al.,* 2016). The main source of chloride in lake water include salt deposits, untreated sewage, industrial effluents and through agricultural inputs (Sajeev *et al.,* 2020).

Total Hardness, Calcium and Magnesium

Total hardness is caused due to the presence of cations and anions such as calcium and magnesium, carbonates, bicarbonates and chloride in lake water (Kamboj and Kamboj, 2019). Calcium levels in water help in osmoregulation under stressful conditions and are important for egg and larvae development in fishes. A calcium concentration of >400 mg/L is detrimental to fish and crustaceans (Stone *et al.,* 2013). Magnesium generally occurs in lesser concentrations than calcium in lakes. The total hardness ranged between 72 mg/L – 510 mg/L in sampled lakes. Calcium and magnesium are necessary for metabolic activities like bone and scale formation in fishes (Bhatnagar and Devi, 2013). A hardness level of 150 mg/L is recommended for the optimum growth and survival of *Labeo rohita* (Rajkumar *et al.,* 2018). The calcium and magnesium varied among lakes and recorded as 14.43 mg/L - 160.32 mg/L and 3.72 mg/L - 42.41 mg/L, respectively (Figure 9.11).

Nitrate and Ortho-phosphate

Nitrate and phosphate are the limiting nutrients in aquatic ecosystem. Nitrate ranged between 0.118 mg/L - 4.283 mg/L among lakes of Bangalore (Figure 9.12).

The increased levels of nitrogen and phosphorus in water occurs due to the discharge of untreated domestic wastes, industrial effluents and agricultural runoff (Pinto *et al.,* 2020). In the present study, ortho-phosphate ranged between 0.06 mg/L - 4.41 mg/L (Figure 9.12). The increase of phosphorus levels in freshwater are due to domestic, agricultural and industrial inputs (external P loading) along with internal P loading from the bottom sediments (Calijuri *et al.,* 2002). The extensive use of detergents and sewage are the major sources of phosphates in lake

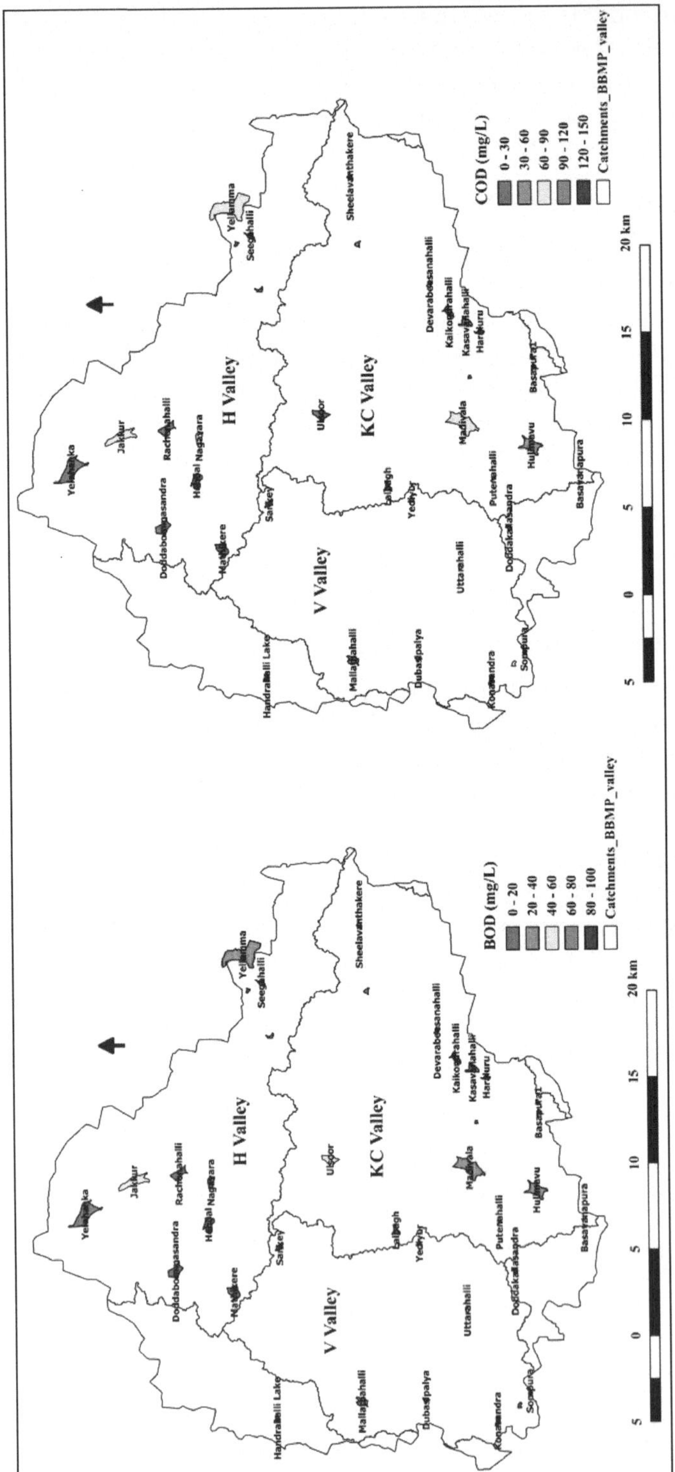

Figure 9.9: Concentration of BOD and COD in Lakes of Bangalore.

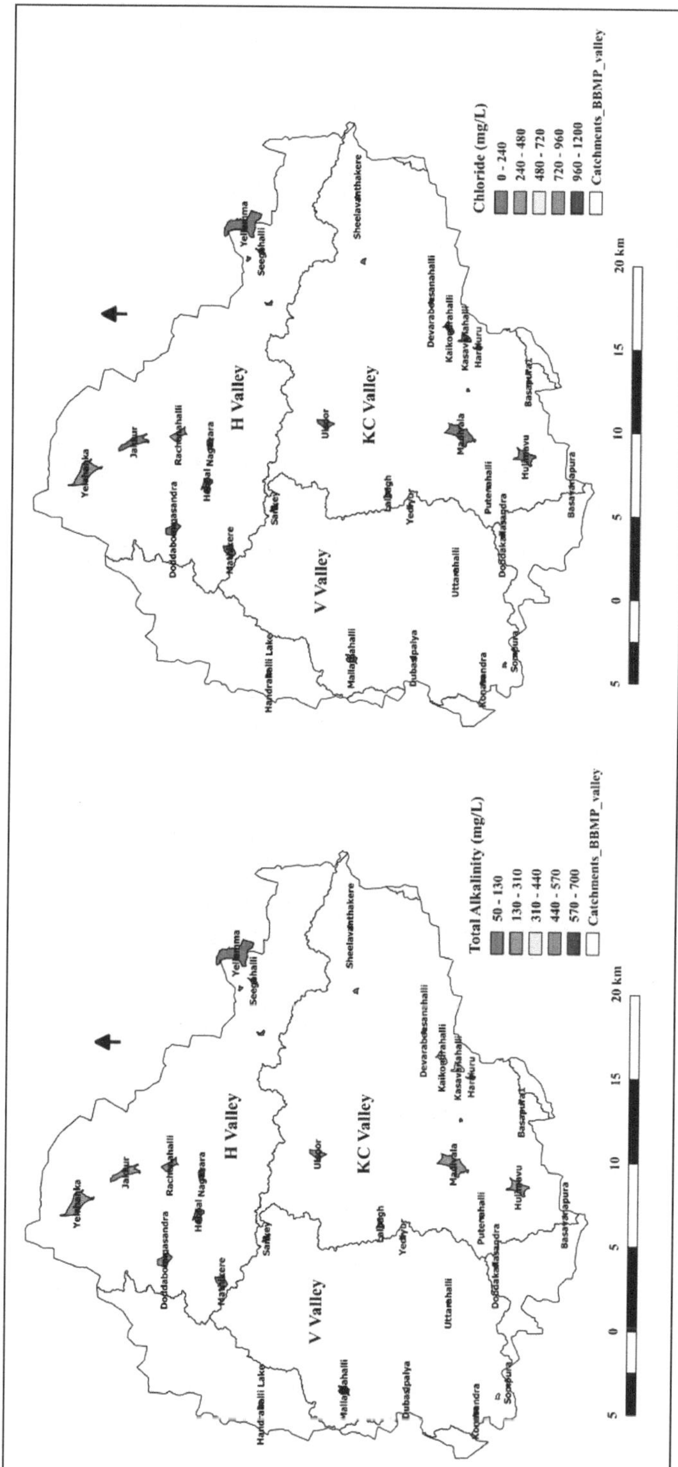

Figure 9.10: Variation of Total Alkalinity and Chloride in Lakes of Bangalore.

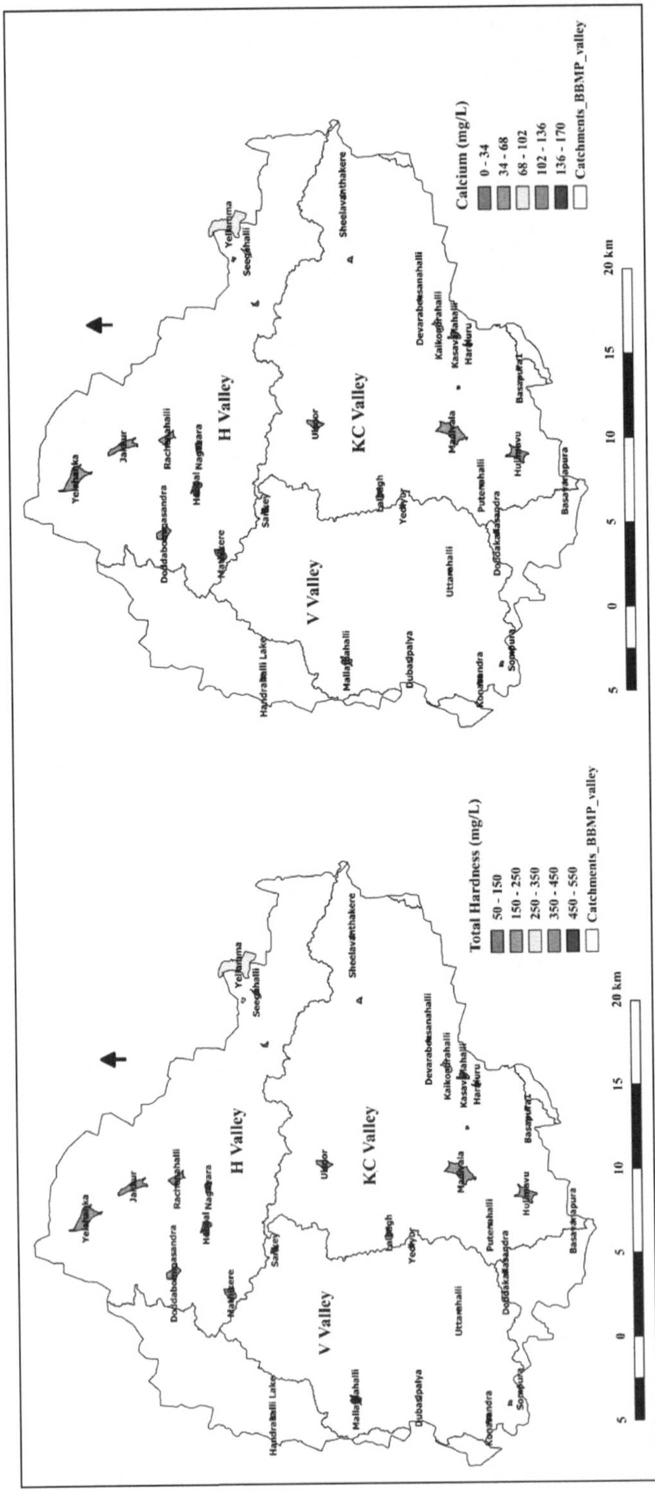

Figure 9.11: The Total Hardness and Calcium Levels in Lakes of Bangalore.

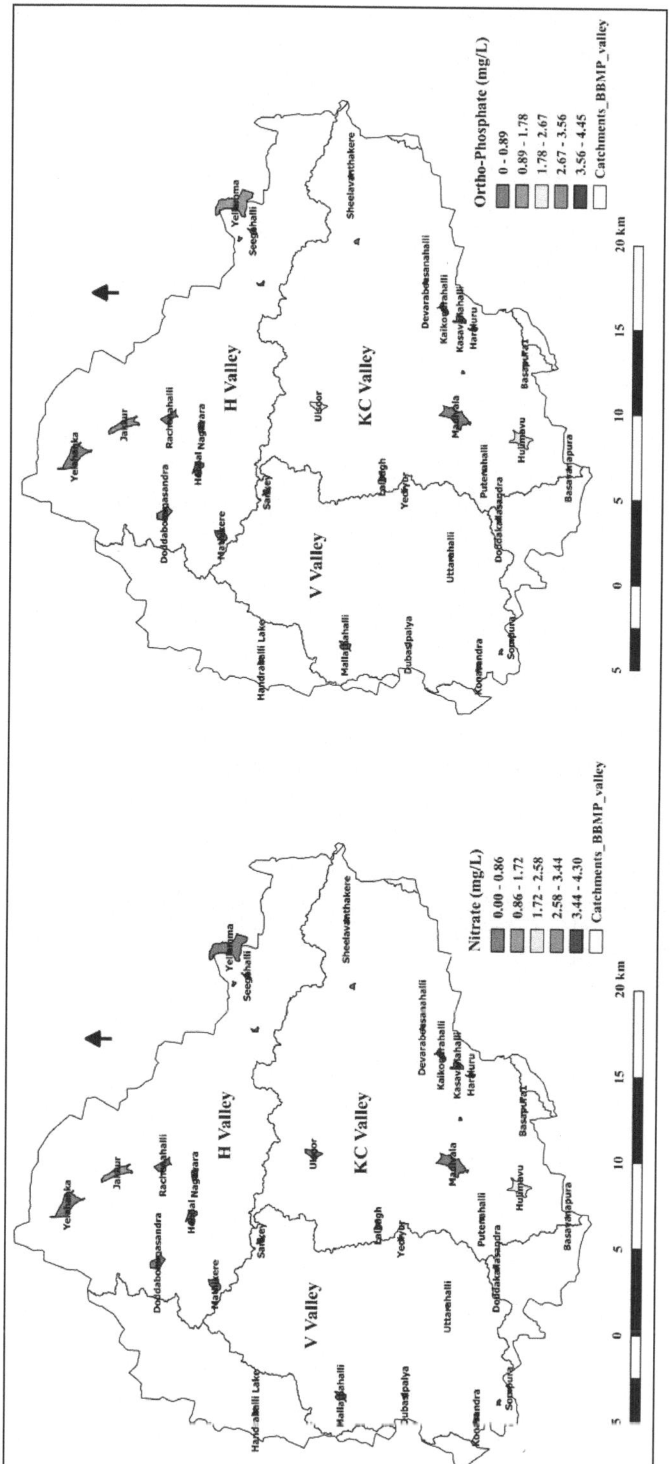

Figure 9.12: Variation of Nitrate and Ortho-phosphate in Lakes of Bangalore.

water (Abdusalam *et al.*, 2019). Water rich in phosphorus favors the growth and proliferation of algae, which results in the eutrophication in water bodies (Rocha *et al.*, 2015). Detergent effluents can induce several toxicological effects in *Clarias gariepinus* (Ogundiran *et al.*, 2010; Nkpondion *et al.*, 2016).

Factors Affecting Lake Water Quality

Principal component analysis (PCA) was applied to the normalized data sets (*i.e.*, 15 physico-chemical parameters of 36 lakes) to identify underlying factors affecting water quality of lakes. The PC loadings of > 0.75, 0.75 - 0.50 and 0.50 - 0.30 were classified as 'strong', 'moderate' and 'weak' respectively (Liu *et al.*, 2003). The principal components (PCs) with eigenvalues greater than one (>1) were considered significant to explain the whole dataset.

In the present study, four principal components were obtained with eigenvalues greater than 1, that explained 81.4 per cent of the total variance in the water quality dataset (Table 9.3). The first component, PC1 accounted for about 47.5 per cent of the total variance in the water quality data set. PC1 has positive loadings on water temperature (0.50), TDS (0.94), EC (0.97), total hardness (0.90), calcium (0.93), magnesium (0.53), chloride (0.88), turbidity (0.76), BOD (0.83) and COD (0.76). This factor is attributed to ionic and organic pollutants, which highlights of industrial and domestic discharges.

Table 9.3: Loadings of Environmental Variables on Principal Components for Water Quality Datasets

	PC 1	*PC 2*	*PC 3*	*PC 4*
Water Temperature	0.50	0.52	-0.02	-0.16
TDS	0.94	-0.10	0.08	0.19
EC	0.97	0.02	-0.04	0.15
pH	0.15	0.57	0.61	0.07
Total Hardness	0.90	-0.36	0.03	0.11
Calcium	0.93	-0.17	-0.08	0.09
Magnesium	0.53	-0.62	0.26	0.12
Chloride	0.88	0.08	-0.16	0.25
Total Alkalinity	0.41	-0.73	0.19	0.00
Turbidity	0.76	0.40	-0.29	-0.15
DO	0.06	0.51	0.18	0.73
BOD	0.83	0.23	-0.08	-0.35
COD	0.76	0.30	-0.33	-0.18
OP	0.45	-0.34	0.53	-0.36
Nitrate	0.28	0.58	0.53	-0.22
Eigenvalue	7.1	2.7	1.3	1.1
Per cent variance	47.5	18.0	8.7	7.2

PC2 accounted for about 18.0 per cent of the total variance with positive loadings of water temperature (0.52), pH (0.57), nitrate (0.58) and DO (0.51) and negative loadings of magnesium (-0.62) and alkalinity (-0.73). This factor corresponds to organic pollution and relates to the productivity of freshwater ecosystem. The PC3 explained 8.7 per cent of the total variance with positive loadings of ortho-phosphate (0.53), nitrate (0.53) and pH (0.61). This factor represents the varied sources of nutrient pollution due to domestic sources, agricultural activities (fertilizer input), industrial discharges and urbanization (urban wastewater). PC4 explained 7.2 per cent of the total variance and is related to DO (0.73) which suggests that lakes had enough amount of dissolved oxygen to sustain aquatic life. A healthy aquatic ecosystem holds DO level of 4 - 6 mg/L (Avvannavar and Shrihari, 2008). DO varies with time, season, rate of photosynthesis, decomposition and respiration activities in lake water. The untreated sewage flowing into lakes alters water chemistry by increasing the ionic, organic and nutrient contents.

WQI Status of Lakes in Bangalore

WQI is computed considering ten water quality parameters - pH, electrical conductivity, total dissolved solids, dissolved oxygen, total hardness, calcium, magnesium, chloride, total alkalinity and nitrate. The water quality condition is described on the basis of WQI value in the range of 0-25, 26-50, 51-75, 76-100 and >100 which corresponds to excellent, good, poor, very poor and unsuitable respectively. Overall, Water quality index (WQI) for the lake water samples were found in the range of 38 to 122 (Figure 9.13).

In the current study on lakes of Bangalore, about 5 per cent of lakes (Lalbagh and Yediyur) fell under good category while 31 per cent lakes fell under poor category. Majority of lakes (about 56 per cent) were classified under very poor category. Lakes namely Basavanapura, Mallathahalli and Sheelavanthakere which fell under the category of unsuitable (8 per cent), is not appropriate for fish culture. The water quality of lakes is deteriorating as evident from WQI status of lakes, mainly due to the inflow of sewage water from residential and commercial complexes, agricultural run-off, anthropogenic activities, untreated effluents from industries and factories, dumping of solid wastes into lakes and lack of proper sanitation.

In earlier studies, WQI values of Sankey tank water belonged to good water class whereas Mallathahallilake fell under poor water category (Ravikumar *et al.*, 2013). WQI values of Hebballake fell under very poor category (Sudarshan *et al.*, 2019). Earlier study has reported WQI of 3 lakes in Haryana of > 100, indicating that water is unsuitable for drinking, outdoor bathing and other uses (Kumar *et al.*, 2018). WQI status of Pariyejlake is poor and unfit for human consumption (Thakor *et al.*, 2011). Anthropogenic stress caused the deterioration of water quality of Dal lake as evident from WQI *i.e.*, polluted and unfit for human consumption (Ahmad *et al.*, 2020).

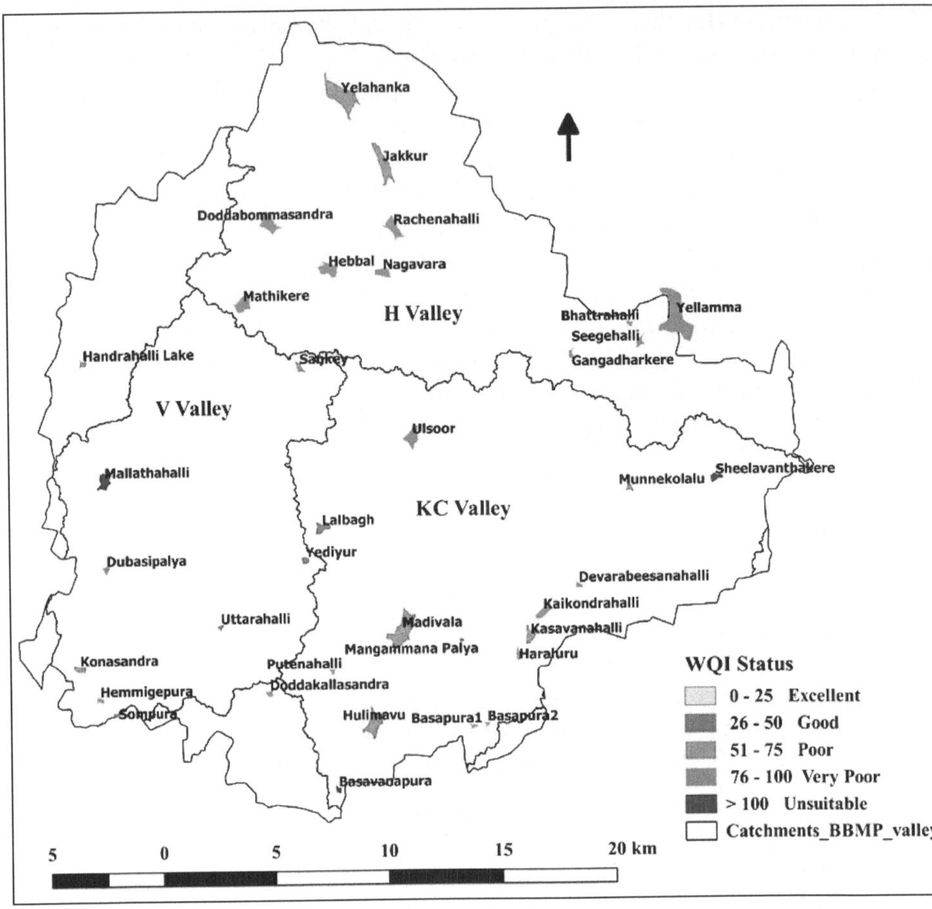

Figure 9.13: WQI Status of Lakes of Bangalore.

Water quality parameters such as temperature, suspended solids, pH, DO, ammonia, nitrite, carbon-dioxide and alkalinity play a crucial role in the growth, reproduction and survival of fish (Okoliegbe *et al.,* 2020). The diversity and distribution of ichthyofaunais governed by pH, turbidity and electrical conductivity (Shetty *et al.,* 2015). The climate change with a decline in rainfall and delayed monsoon, affects temperature of freshwater ecosystems and the breeding behavior/ ecology of fishes (Ninawe *et al.,* 2018). The growth rate of *Oreochromis niloticus* increased at higher temperature and DO whereas the growth rate declined under increased pH, conductivity and ammonia levels (Makori *et al.,* 2017). The sewage and industrial effluents let into water bodies increases the levels of total suspended solids, total dissolved solids, COD and BOD (Ramanujam *et al.,* 2014). The detergent effluents in lake water induce severe damage to gills, skin, kidney, heart, liver and brain of fish (Nkpondion *et al.,* 2016).Inorganic pollution from industries is another main threat to ichthyofauna (Rao *et al.,* 2014). The waterdepth, pressure, turbulence, temperature, light and turbidity are important for fish culture (Priyamvada *et al.*, 2013), which necessitates maintaining water quality to sustain fish production.

In 1970s, about 55 fish species were recorded in Bangalore. At present, there is a massive reduction in ichthyofauna diversity in Bangalore as the lakes are under threat due to water pollution,urbanization, encroachment of lakes, habitat loss or habitat degradation, invasion of exotic species, agricultural practices, climate change,flood, drought, over-harvesting of fish resources and loss of interconnectivity among lakes. The anthropogenic activities, water pollution, eutrophication, habitat degradation, overexploitation, hydrologic alterations, flow modification, damconstruction and climate change are the major threats to fish biodiversity (Borah and Das, 2020; Bhakta *et al.*, 2019; Gupta *et al.*, 2015; Vijaylaxmi *et al.*, 2010). Introduction of exotic species and adopting destructive fishing method such as dynamite or poisoning would cause serious threat to native ichthyofauna (Bose *et al.*, 2019). Urbanization, pollution and water abstraction for irrigation and power generation impose threat to fish diversity (Kumar Sarkar *et al.*, 2013). Hence, it is essential to regulate the sustained inflow of untreated sewage and industrial effluents into lakes to maintain the integrity of aquatic ecosystems and sustain fish diversity. The nutrient removal in lakes (bioremediation) can be achieved in a cost effective way through the integration of conventional water treatment methods with the constructed wetlands (macrophytes) and algal pond (Ramachandra *et al.*, 2018).In order to increase the ichthyofauna diversity, there is a need to avoid habitat destruction, control pollution sources, should ban the introduction of invasive species, harvesting of fish during the spawning period and the harvesting of juveniles. Environmental awareness through awareness programs would help in educating public on the impacts of water pollution on freshwater fish and also the role of healthy ecosystem in supporting people's livelihood. Regular workshops and awareness programs on wetland goods and services would also help in the conservation of aquatic biodiversity.

Conclusion

The current study recorded the ichthyofauna diversity and assessed the water quality of 36 lakes in Bangalore district, Karnataka. The order Cypriniformesis dominant in Bangalore lakes. The rapiddecline inichthyofauna diversity is noticed in Bangalore lakes due to water pollution, habitat degradation,overfishing,invasion of exotic species and climate change. A decline in the population of native species is evident due to the invasion of exotic species. The lakes of Bangalore are polluted due to the sustained inflow of raw sewage and partially treated wastewater. In the present study, PCA helped in understanding the vital physico-chemical parameters that caused changes in lake water quality. PCA of the water quality data evolved four PCs, explaining about 81.4 per cent of the total variance. The main factors affecting lake water quality are water temperature, TDS, EC, total hardness, calcium, magnesium, chloride, turbidity, BOD and COD. These factors correspond to ionic and organic pollution of lakes mainly due to the sustained discharge of untreated or partially treated domestic sewage and industrial wastewater. The water quality index (WQI) for the 36 monitored lakes, revealed that about 56 per cent of lakes fell under very poor water quality category. Lakes such as Basavanapura, Mallathahalli and Sheelavanthakere are polluted and are with poor fish diversity and yield. This necessitates rejuvenation of lakes and arresting anthropogenic pollution sources to protect the aquatic life in lakes of Bangalore. A regular monitoring of lakes and its biodiversity is necessary to raise awareness among the community and conserve the fragile productive ecosystems.

Acknowledgments

We thank The ENVIS Division, The Ministry of Environment, Forests and Climate Change, Government of India and Indian Institute of Science (IISc) for the sustained support to ecological research.

REFERENCES

Abdusalam, Y.H., Sujaul, I.M., Karim, M.A., Salah, M.G., Ali, M.I., and Ramli, N.I. (2019). Assessment of water quality in the vicinity of Chini lake, Malaysia. *Bangladesh Journal of Botany, 48*(4), 1037-1046.

Ahmad, T., Gupta, G., Sharma, A., Kaur, B., Alsahli, A.A., and Ahmad, P. (2020). Multivariate statistical approach to study spatiotemporal variations in water quality of aHimalayan urban freshwater lake. *Water, 12*(9), 2-15.doi: 10.3390/ w12092365

Alam, M.S., and Aslam, R. (2020). Extrusion for the Production of Functional Foods and Ingredients. *Innovative Food Processing Technologies*, 22-35.

APHA, AWWA, WEF. (2005). Standard methods for the examination of water and wastewater. 21st Edition. D.C., Washington: American Public Health Association.

Atkore, V.M., Sivakumar, K., and Johnsingh, A.J.T. (2011). Patterns of diversity and conservation status of freshwater fishes in the tributaries of River Ramganga in the Shiwaliks of the Western Himalaya. *Current science*, 731-736.

Avvannavar, S. M., and Shrihari, S. (2008). Evaluation of water quality index for drinking purposes for river Netravathi, Mangalore, South India. *Environmental Monitoring and Assessment, 143*(1-3), 279-290. DOI 10.1007/s10661-007-9977-7

Bahar, M.M., Ohmori, H., and Yamamuro, M. (2008). Relationship between river water quality and land use in a small river basin running through the urbanizing area of Central Japan. *Limnology, 9*(1), 19-26. DOI 10.1007/s10201-007-0227-z

Barakat, A., El Baghdadi, M., Rais, J., Aghezzaf, B., and Slassi, M. (2016). Assessment of spatial and seasonal water quality variation of OumErRbia River (Morocco) using multivariate statistical techniques. *International Soil and*

Water Conservation Research, 4(4), 284-292.http: //dx.doi.org/10.1016/j.iswcr.2016.11.002

Barik, N.K. (2017). Freshwater fish for nutrition security in India: Evidence from FAO data. *Aquaculture Reports, 7*, 1-6. http: //dx.doi.org/10.1016/j.aqrep.2017.04.001

Bassi, N., Kumar, M.D., Sharma, A., and Pardha-Saradhi, P. (2014). Status of wetlands in India: A review of extent, ecosystem benefits, threats and management strategies. *Journal of Hydrology: Regional Studies, 2*, 1-19.http: //dx.doi.org/10.1016/j.ejrh.2014.07.001

Bhakta, D., Meetei, W.A., Vaisakh, G., Kamble, S., Das, S.K., and Das, B.K. (2019). Finfish diversity of Narmada estuary in Gujarat of India. *In Proceedings of the Zoological Society, 72*(3), 257–262. https: //doi.org/10.1007/s12595-018-0263-1

Bhatnagar, A., and Devi, P. (2013). Water quality guidelines for the management of pond fish culture. *International Journal of Environmental Sciences, 3*(6), 1980 - 2009.

Bora, M., and Goswami, D.C. (2017). Water quality assessment in terms of water quality index (WQI): Case study of the Kolong River, Assam, India. *Applied Water Science, 7*(6), 3125-3135.DOI 10.1007/s13201-016-0451-y

Borah, D.K., and Das, J. (2020). Ichthyofaunal diversity of Jinari river in Goalpara, Assam, India. *CIBTech Journal of Zoology. 9*, 30-35.

Bose, R., Bose, A.K., Das, A.K., Parashar, A., and Roy, K. (2019). Fish diversity and limnological parameters influencing fish assemblage pattern in Chambal River Basin of Madhya Pradesh, India. *Proceedings of the National Academy of Sciences, India Section B: Biological Sciences, 89*(2), 461-473.https: //doi.org/10.1007/s40011-017-0958-5

Calijuri, M.C., Dos Santos, A.C.A., and Jati, S. (2002). Temporal changes in the phytoplankton community structure in a tropical and eutrophic reservoir (Barra Bonita, SP–Brazil). *Journal of Plankton Research, 24*(7), 617-634.

Chabuk, A., Al-Madhlom, Q., Al-Maliki, A., Al-Ansari, N., Hussain, H.M., and Laue, J. (2020). Water quality assessment along Tigris River (Iraq) using water quality index (WQI) and GIS software. *Arabian Journal of Geosciences, 13*(14), 1-23.

Chatterjee, M., Ghosh, P., Ramdas, L., and Chakrabarti, R. (2015). Isotopic and geochemical characterization of invader tilapia fishes from water bodies of West Bengal and Karnataka, India. *Environmental Monitoring and Assessment, 187*(11), 1-12.DOI 10.1007/s10661-015-4929-0

Chattopadhyay, N.R. (2017). Reproductive cycle, maturation, and spawning: A practical guide for hatcheries. *Induced Fish Breeding*, 15-42.

Davraz, A., Varol, S., Sener, E., Sener, S., Aksever, F., Kýrkan, B., and Tokgözlü, A. (2019). Assessment of water quality and hydrogeochemical processes of Salda alkaline lake (Burdur, Turkey). *Environmental Monitoring and Assessment, 191*(11), 1-18.https: //doi.org/10.1007/s10661-019-7889-y

Devi, W.S., Singh, K., and Meitei, N. (2015). Assessment of water quality index of Nambol River, Manipur, India. *Universal Journal of Environmental Research and Technology, 5*(3), 165-172.

Diamantini, E., Lutz, S.R., Mallucci, S., Majone, B., Merz, R., and Bellin, A. (2018). Driver detection of water quality trends in three large European river basins. *Science of the Total Environment, 612*, 49-62.http: //dx.doi.org/10.1016/j.scitotenv.2017.08.172

Ekubo, A.T., and Abowei, J.F.N. (2011). Review of some water quality management principles in culture fisheries. *Research Journal of Applied Sciences, Engineering and Technology, 3*(12), 1342-1357.

El-Serehy, H.A., Abdallah, H.S., Al-Misned, F.A., Al-Farraj, S.A., and Al-Rasheid, K.A. (2018). Assessing water quality and classifying trophic status for scientifically based managing the water resources of the Lake Timsah, the lake with salinity stratification along the Suez Canal. *Saudi Journal of Biological Sciences, 25*(7), 1247-1256. https: //doi.org/10.1016/j.sjbs.2018.05.022

Gupta, N., Sivakumar, K., Mathur, V.B., and Chadwick, M.A. (2015). Terrestrial protected areas and managed reaches conserve threatened freshwater fish in Uttarakhand, India. *Parks, 21*(1), 89-101.

Haidary, A., Amiri, B.J., Adamowski, J., Fohrer, N. and Nakane, K., 2013. Assessing the impacts of four land use types on the water quality of wetlands in Japan. *Water Resources Management, 27*(7), 2217-2229.DOI 10.1007/s11269-013-0284-5

http: //fishbase.org

https: //www.iucnredlist.org/

Huang, J., Zhan, J., Yan, H., Wu, F., and Deng, X. (2013). Evaluation of the impacts of land use on water quality: a case study in the Chaohu Lake basin. *The Scientific World Journal, Article*, 1-7.http: //dx.doi.org/10.1155/2013/329187

Islam, M.M., Lenz, O.K., Azad, A.K., Ara, M.H., Rahman, M., and Hassan, N. (2017). Assessment of spatio-temporal variations in water quality of Shailmari River, Khulna (Bangladesh) using multivariate statistical techniques. *Journal of Geoscience and Environment Protection, 5*(1), 1-26. http: //dx.doi.org/10.4236/gep.2017.51001

Jayaram K.C. (1999). The freshwater fishes of the Indian Region. Narendra Publishing House, Delhi- 551.

Jayasankar, P. (2018). Present status of freshwater aquaculture in India-A review. *Indian Journal of Fisheries, 65*(4), 157-165.

Jindal, R., and Sharma, C. (2011). Studies on water quality of Sutlej River around Ludhiana with reference to physico-chemical parameters. *Environmental*

Monitoring and Assessment, 174(1), 417-425. DOI 10.1007/s10661-010-1466-8

Kamble, P., and Ganesh, C.B. (2016). Ichthyofaunal diversity status in Kelageri Lake, Dharwad, Karnataka state.*International Journal of Fisheries and Aquatic Studies, 4*(4), 343-346.

Kamboj, N., and Kamboj, V. (2019). Water quality assessment using overall index of pollution in riverbed-mining area of Ganga-River Haridwar, India. *Water Science, 33*(1), 65-74.https: //doi.org/10.1080/11104929.2019.1626631

Kumar Sarkar, U., Kumar Pathak, A., Kumar Tyagi, L., Mohan Srivastava, S., Prakash Singh, S., and Kumar Dubey, V. (2013). Biodiversity of freshwater fish of a protected river in India: comparison with unprotected habitat. *Revista de BiologiaTropical, 61*(1), 161-172.

Kumar, A.B. (2000). Exotic fishes and freshwater fish diversity. *Zoos' Print Journal, 15*(11), 363-367.

Kumar, D., Karthik, M., and Rajakumar, R. (2017). Study of seasonal water quality assessment and fish pond conservation in Thanjavur, Tamil Nadu, India.*Journal of Entomology and Zoology Studies, 5*, 1232-1238.

Kumar, R., Grover, A.S., and Wats, M. (2018). Assessment of water quality status of lakes in Haryana, India. *International Journal of Recent Scientific Research, 9*(7), 27831-27835.

Liu, C.W., Lin, K.H., and Kuo, Y.M. (2003). Application of factor analysis in the assessment of groundwater quality in a Blackfoot disease area in Taiwan. *Science of the Total Environment, 313*, 77-89.

Liu, J., Zhang, D., Tang, Q., Xu, H., Huang, S., Shang, D., and Liu, R. (2021). Water quality assessment and source identification of the Shuangji River (China) using multivariate statistical methods. *PloS One, 16*(1), 1-19.

Mahapatra, B.K., Sarkar, U.K., and Lakra, W.S. (2014). A Review on status, potentials, threats and challenges of the fish biodiversity of West Bengal. *Journal of Biodiversity, Bioprospecting and Development, 2*(1), 1-10.doi: 10.4172/2376-0214.1000140

Makori, A.J., Abuom, P.O., Kapiyo, R., Anyona, D.N., and Dida, G.O. (2017). Effects of water physico-chemical parameters on tilapia (*Oreochromis niloticus*) growth in earthen ponds in Teso North Sub-County, Busia County. *Fisheries and Aquatic Sciences, 20*(1), 1-10.DOI 10.1186/s41240-017-0075-7

Mamun, M., Kim, J.Y., and An, K.G. (2021). Multivariate statistical analysis of water quality and trophic state in an artificial dam reservoir. *Water, 13*(2), (1-18).

Mishra, A. (2010). Assessment of water quality using principal component analysis: a case study of the river Ganges. *Journal of Water Chemistry and Technology, 32*(4), 227-234.

Mohamed, A.R.M., and Al-Wan, S.M. (2020). Biological aspects of an invasive species of Oreochromis niloticus in the Garmat Ali River, Basrah, Iraq. *Journal of Agriculture and Veterinary Science, 13*(2), 15-26.

Naskar, M., Roy, K., Karnatak, G., Nandy, S.K., and Roy, A. (2018). Quantifying climate change induced threats to wetland fisheries: a stakeholder-driven approach. *Environment, Development and Sustainability, 20*(6), 2811-2830. https: //doi.org/10.1007/s10668-017-0018-6

Nayaka, B.M.S. (2018). Ichthyofaunal diversity of Mallasandra Lake of Tumakuru, Karnataka State, India. *National Journal of Multidisciplinary Research and Development, 3*(2), 15-17.

Ninawe, A.S., Indulkar, S.T., and Amin, A. (2018). Impact of climate change on fisheries. *In Biotechnology for Sustainable Agriculture: Emerging Approaches and Strategies, Elsevier*, 257-280. https: //doi.org/10.1016/B978-0-12-812160-3.00009-X

Nkpondion, N.N., Ugwumba, O.A., and Esenowo, I.K. (2016). The toxicity effect of detergent on enzymatic and protein activities of African Mud Catfish (*Clarias gariepinus*). *Journal of Environmental and Analytical Toxicology, 6*(361), 2161-0525.

Ogundiran, M.A., Fawole, O.O., Adewoye, S.O., and Ayandiran, T.A. (2010). Toxicological impact of detergent effluent on juvenile of African Catfish (Clarias gariepinus)(Buchell 1822). *Agriculture and Biology Journal of North America, 1*(3), 330-342.

Okoliegbe, I.L., Ariole, C.N., and Okpokwasili, G.C. (2020). Physico-chemical properties of concrete pond water used for *Clarias gariepinus* aquaculture. *Nature and Science, 18*(7), 48-55. doi: 10.7537/mars nsj180720.07.

Pinto, F.P., Tormam, M.F., Bork, C.K., Guedes, H.A.S., and Silva, L.B.P.D. (2020). Seasonal assessment of water quality parameters in Mirim Lagoon, Rio Grande do Sul State, Brazil. *Anais da Academia Brasileira de Ciências, 92*(3), 1-15.

Prasad, S.N., Ramachandra, T.V., Ahalya, N., Sengupta, T., Kumar, A., Tiwari, A.K., Vijayan, V.S., and Vijayan, L. (2002). Conservation of wetlands of India-A review. *Tropical Ecology, 43*(1), 173-186.

Priyamvada, D., Sirisha D., and Gandhi N. (2013). Study on the quality of water and soil from fish pond in around Bhimavaram West Godavari district, A.P., INDIA. *International Research Journal of Environment Sciences, 2*(1), 58-62.

Qureshimatva, U.M., Maurya, R.R., Gamit, S.B., Patel, R.D., and Solanki, H.A., 2015. Determination of physico-chemical parameters and water quality index (WQI) of Chandlodia Lake, Ahmedabad, Gujarat, India. *Journal of Environmental and Analytical Toxicology, 5*(4), 1-6.DOI: 10.4172/2161-0525.1000288

Rajbongshi, M.K., Das, J., and Dutta, R.K. (2016). Water quality assessment of capture and culture fishery in Barpeta district, Assam, India. *International Journal of Fisheries and Aquatic Studies, 4*(5), 516-520.

Rajkumar, K., Ojha, M.L., Saini, V.P., and Sharma, S.K. (2018). Effect of water hardness on survival and growth of *Labeo rohita* (Hamilton) fry. *Journal of Entomology and Zoology Studies*, *6*(5), 2337-2341.

Ramachandra T.V., Sincy V., and Asulabha K.S. (2020b). Efficacy of rejuvenation of lakes in Bengaluru, India. *Green Chemistry and Technology Letters*, *6*(1), 14-26. doi.org/10.18510/gctl.2020.613

Ramachandra, T.V., and Aithal, B.H. (2016). Bengaluru's reality: Towards unlivable status with unplanned urban trajectory. *Current Science, 110*(12), 2207-2208

Ramachandra, T.V., Asulabha, K.S., and Lone, A.A. (2014). Nutrient enrichment and proliferation of invasive macrophytes in urban lakes. *Journal of Biodiversity*, *5*(1-2), 33-44.

Ramachandra, T.V., Sincy, V., Asulabha, K.S., Mahapatra, D.M., Bhat, S.P. and Aithal, B.H. (2018). Optimal treatment of domestic wastewater through constructed wetlands. *Journal of Biodiversity, 9*(1-2), 81-102. DOI: 10.31901/24566543.2018/09.1-2.077

Ramachandra, T.V., Sudarshan, P., Vinay, S., Asulabha, K.S., and Varghese, S. (2020a). Nutrient and heavy metal composition in select biotic and abiotic components of Varthur wetlands, Bangalore, India. *SN Applied Sciences*, *2*(8), 1-14.

Ramachandra, T. V., Vinay, S., Asulabha, K. S., Sincy V., Bhat, S. P., Mahapatra, D.M., and Aithal, B. H. (2017). Rejuvenation blueprint for lakes in Vrishabhavathi Valley.*ENVIS Technical Report 122*, Energy and Wetlands Research Group, CES, Indian Institute of Science, Bangalore, 2017.

Ramanujam, M.E., Devi, K.R., and Indra, T.J. (2014). Ichthyofaunal diversity of the Adyar Wetland complex, Chennai, Tamil Nadu, Southern India. *Journal ofThreatened Taxa*, *6*(4): 5613–5635; http: //dx.doi.org/10.11609/JoTT. o2905.5613-35

Ramulu, N.K., and Benarjee, G. (2013). Fish species diversity of Nagaram Tank of Warangal, Andhra Pradesh. *Journal of Environmental Science, Toxicology and Food Technology*, *3*(4), 14-18.

Rao, J.C.S., Raju, C.S., and Simhachalam, G. (2014). Biodiversity and conservation status of fishes of River Sarada, Visakhapatnam District, Andhra Pradesh, India. *Research Journal of Animal, Veterinary and Fishery Sciences*, *2*(2), 1-8.

Ravikumar, P., Mehmood, M.A., and Somashekar, R.K. (2013). Water quality index to determine the surface water quality of Sankey Tank and Mallathahalli Lake, Bangalore Urban District, Karnataka, India. *Applied Water Science*, *3*(1), 247-261.

Rocha, F.C., Andrade, E.M., and Lopes, F.B. (2015). Water quality index calculated from biological, physical and chemical attributes. *Environmental Monitoring and Assessment*, *187*(1), 1-15. DOI 10.1007/s10661-014-4163-1

Rouse, R.D. (1979). Water quality management in pond fish culture. Research and Development Series no. 22, Project: AID/DSAN-G 0039, Auburn University, Alabama.

Sajeev, S., Sekar, S., Kumar, B., Senapathi, V., Chung, S.Y., and Gopalakrishnan, G. (2020). Variations of water quality deterioration based on GIS techniques in surface and groundwater resources in and around Vembanad Lake, Kerala, India. *Geochemistry, 80*(4), 1-20. https: //doi.org/10.1016/j.chemer.2020.125626

Sandilyan, S. (2016). Occurrence of ornamental fishes: a looming danger for inland fish diversity of India. *Current Science*, 2099-2104.

Saraswathy, R., Muralidhar, M., Sundaray, J. K., Lalitha, N., and Kumararaja, P. (2015). Water quality management in fish hatchery and grow-out systems. *In Advances in Marine and Brackishwater Aquaculture*, 217-225, Springer, New Delhi.

Sarkar, R., Ghosh, A.R., and Mondal, N.K. (2020). Comparative study on physico-chemical status and diversity of macrophytes and zooplanktons of two urban ponds of Chandannagar, WB, India. *Applied Water Science, 10*(2), 1-8.

Sarma, D., Akhtar, M.S., Das, P., Das, P., Shahi, N., Ciji, A., Mahanta, P.C., Yengkokpam, S., and Debnath, D. (2013). Nutritional quality in terms of amino acid and fatty acid of five coldwater fish species: implications to human health. *National Academy Science Letters, 36*(4), 385-391.DOI 10.1007/s40009-013-0151-1

Sharma, D., and Kansal, A. (2011). Water quality analysis of River Yamuna using water quality index in the national capital territory, India (2000–2009). *Applied Water Science, 1*(3), 147-157. DOI 10.1007/s13201-011-0011-4

Shetty, A., Venkateshwarlu, M., and Muralidharan, M. (2015). Effect of water quality on the composition of fish communities in three coastal rivers of Karnataka, India. *International Journal of Aquatic Biology, 3*(1), 42-51.

Sincy, V., Asulabha, K.S., Vinay, S., Vishnu, D.M., Srikanth, N., Subashchandran, M.D., and Ramachandra, T.V. (2016). Ecological status of lotic ecosystem in kans and non-kans of central Western Ghats. Proceedings Lake 2016: *Conservation and Sustainable Management of Ecologically Sensitive Regions in Western Ghats*, 195-208.

Singh, A.K., Kumar, D., Srivastava, S.C., Ansari, A., Jena, J.K., and Sarkar, U.K. (2013). Invasion and impacts of alien fish species in the Ganga River, India. *Aquatic Ecosystem Health and Management, 16*(4), 408-414.

Singh, G., Patel, N., Jindal, T., Srivastava, P., and Bhowmik, A. (2020). Assessment of spatial and temporal variations in water quality by the application of multivariate statistical methods in the Kali River, Uttar Pradesh, India. *Environmental Monitoring and Assessment, 192*, 1-26.https: //doi.org/10.1007/s10661-020-08307-0

Singh, K.P., Malik, A., Mohan, D., and Sinha, S. (2004). Multivariate statistical techniques for the evaluation of spatial and temporal variations in water quality of Gomti River (India) - A case study. *Water Research, 38*(18), 3980-3992.

Siriwardana, C., Cooray, A.T., Liyanage, S.S., and Koliyabandara, S.M.P.A. (2019). Seasonal and spatial variation of dissolved oxygen and nutrients in Padaviya Reservoir, Sri Lanka. *Journal of Chemistry*, 1-11. https: //doi. org/10.1155/2019/5405016

Stone, N.M., Shelton, J.L., Haggard, B.E., and Thomforde, H.K. (2013). Interpretation of water analysis reports for fish culture.*Southern Regional Aquaculture Center*, 1-12.

Sudarshan, P., Mahesh, M.K., and Ramachandra, T.V. (2019). Assessment of seasonal variation in water quality and water quality index (WQI) of Hebbal Lake, Bangalore, India. *Environment and Ecology, 37*(1B), 309-317.

Talwar, P.K. and Jhingran, A.G. (1991). Inland fishes of India and adjacent countries. Oxford and IBH Publishing Co. Pvt. Ltd., New Delhi.

Thakor, F.J., Bhoi, D.K., Dabhi, H.R., Pandya, S.N., Chauhan, N.B., and Chauhan, N.B. (2011). Water quality index (WQI) of Pariyej Lake Dist. Kheda-Gujarat. *Current World Environment, 6*(2), 225-231.

Thirumala, S., Kiran, B. R., and Kantaraj, G. S. (2011). Fish diversity in relation to physico-chemical characteristics of Bhadra reservoir of Karnataka, India. *Advances in Applied Science Research, 2*(5), 34-47.

Vega, M., Pardo, R., Barrado, E., and Debán, L. (1998). Assessment of seasonal and polluting effects on the quality of river water by exploratory data analysis. *Water Research, 32*(12), 3581-3592.

Vijaylaxmi, C., Rajshekhar, M., and Vijaykumar, K. (2010). Freshwater fishes distribution and diversity status of Mullameri River, a minor tributary of Bheema River of Gulbarga District, Kamataka. *International Journal of Systems Diology, 2*(2), 1-9.

Wang, J., Liu, G., Liu, H., and Lam, P.K. (2017). Multivariate statistical evaluation of dissolved trace elements and a water quality assessment in the middle reaches of Huaihe River, Anhui, China. *Science of the Total Environment, 583*, 421-431.

Wu, Z., Zhang, D., Cai, Y., Wang, X., Zhang, L., and Chen, Y. (2017). Water quality assessment based on the water quality index method in Lake Poyang: the largest freshwater lake in China. *Scientific Reports, 7*(1), 1-10.

Yýlmaz, E., Koç, C., and Gerasimov, I. (2020). A study on the evaluation of the water quality status for the Büyük Menderes River, Turkey. *Sustainable Water Resources Management, 6*(6), 1-16. https: //doi.org/10.1007/s40899-020-00456-x

Zeinalzadeh, K., and Rezaei, E. (2017). Determining spatial and temporal changes of surface water quality using principal component analysis. *Journal of Hydrology: Regional Studies*, 13, 1-10.http: //dx.doi.org/10.1016/j.ejrh.2017.07.002

Zhang, X., Wang, Q., Liu, Y., Wu, J., and Yu, M. (2011). Application of multivariate statistical techniques in the assessment of water quality in the Southwest New Territories and Kowloon, Hong Kong. *Environmental Monitoring and Assessment, 173*(1), 17-27.DOI 10.1007/s10661-010-1366-y

Chapter 10

Diversity of Phytoplankton in Lakes of Bangalore, Karnataka, India

*K.S. Asulabha[1,4], R. Jaishanker[4], V. Sincy[1,4] and T.V. Ramachandra[1,2,3]**

[1]Energy and Wetlands Research Group [CES TE15], Centre for Ecological Sciences,
[2]Centre for Sustainable Technologies (Astra)
[3]Centre for Infrastructure, Sustainable Transportation and Urban Planning (CiSTUP)
Indian Institute of Science, Bangalore – 560 012, Karnataka
[4]Indian Institute of Information Technology and Management-Kerala (IIITM-K), Thiruvananthapuram, Kerala
**Corresponding author e-mail: tvr@iisc.ac.in*

Phytoplanktons are an essential component in the ecosystem as they provide food (primary producer) to higher trophic levels through the process of photosynthesis with the release of oxygen as a by-product. They act as an indicator of variations in water quality and helps in nutrient removal in lakes. The diversity, density, and distribution of phytoplankton in a water body are dependent on the quality of the habitat (water quality). The present study deals with the assessment of physico-chemical parameters and phytoplankton diversity of lakes in Bangalore. A total of 58 genera of phytoplankton were recorded in Bangalore lakes. The phytoplankton in the studied lakes was represented by five major groups, i.e., Chlorophyta, Bacillariophyta, Cyanobacteria, Charophyta, and Euglenozoa. Cyanobacteriais the most dominant that suppresses the growth of other phytoplankton groups. The primary production evident from the composition of phytoplankton in the lakes depends on the environmental parameters. Canonical Correspondence Analysis (CCA) of the water quality parameters and phytoplankton genera indicated the influence of water quality on phytoplankton composition in monitored lakes of Bangalore. A reduction in phytoplankton diversity is attributed to a decline in the water quality of lakes with increased water pollution. The phytoplankton serves as excellent bioindicators as it responds to environmental stress. Hence, phytoplanktons are being used in the biomonitoring programs of lakes.

***Keywords**: Phytoplankton, Lakes, Water quality, Cyanophyceae, Biomonitoring.*

Introduction

Wetlands play a vital role in providing food and fiber, microclimate regulation, water purification, groundwater recharge, flood protection, erosion prevention, biodiversity conservation, recreation, *etc.* (Ramachandra *et al.*, 2020a; Li *et al.*, 2020). Sustaining these services entails maintaining the ecological integrity of wetlands through regular monitoring of wetlands, which involves assessing the physical, chemical, and biological characteristics (Alves *et al.*, 2018). The spatial and temporal variations in the water quality of freshwater ecosystems are associated with several natural factors and anthropogenic factors (Varol, 2020; Shil *et al.*, 2019). The land use land cover [LULC] of a region influences water quality (Chen and Lu, 2014; Xu *et al.*, 2019). Anthropogenic activities cause water quality deterioration and render the water unfit for drinking, domestic purpose, irrigation, industrial and recreational use. Unplanned rapid urbanization and the consequent disposal of untreated or partially treated wastewater have been deteriorating water resources (Ramachandra *et al.*, 2020b).

Phytoplankton (microalgae) are ubiquitous photosynthetic organisms that occur in water bodies. Phytoplanktons as producers play a key role in nutrient cycling and transformations and the productivity of aquatic ecosystems. Phytoplanktons consist of both eukaryotic (green algae, diatoms, *etc.*) and prokaryotic forms (cyanobacteria). Algae provide food (energy transfer) and dissolved oxygen for consumers at higher trophic levels in an aquatic ecosystem. Phytoplankton has a plethora of applications such as pharmaceuticals, nutraceuticals, feed, and aid in bioremediation. For the past few decades, microalgae have been explored for biofuel production, electricity generation, wastewater treatment, greenhouse gas mitigation, and CO_2 removal (Das *et al.*, 2021; Qu *et al.*, 2019; Xu *et al.*, 2015). The influence of physico-chemical characteristics of water on the phytoplankton density and diversity is demonstrated in numerous studies (Kozak *et al.*, 2020; Afonina and Tashlykova, 2018; Sharma *et al.*, 2016). The distribution and composition of phytoplankton in lakes or wetlands are influenced by factors such as temperature, light, nutrients, water level, and grazing pressure. Land use and land cover (LULC) changesin the wetland catchment altered the physical,chemical, and biological parameters (Ramachandra *et al.*, 2013).Water level fluctuations and turbulence induce disturbances in the physical environment and concentration of nutrients (Adamczuk *et al.*, 2020), altering phytoplankton communities structure (Ji *et al.*, 2017; Yang *et al.*, 2016; Zhao *et al.*, 2020). The sustained inflow of untreated wastewater to waterbodies enhances the nutrient concentration in water leading to eutrophication, which increases the phytoplankton biomass, and it reduces the nutritional value, affecting higher trophic level organisms (Taipale *et al.*, 2019).Multivariate analyses using canonical correspondence analysis (CCA)have established the relationships between phytoplankton taxa and environmental variables (Fetahi *et al.*, 2014; Abd El-Karim, 2014, Tian *et al.*, 2012).

Phytoplankton serves as a valuable bioindicator of water quality changes as they have a short lifespan and respond quickly to environmental changes and pollutants (El-Kassas and Gharib, 2016; Kireta *et al.*, 2012; Lavoie *et al.*, 2012). The species number and cell density of phytoplankton serve as bioindicators of water quality. Monitoring these bioindicators and the physical and chemical factors help in overall water quality assessment (Jiang *et al.*, 2014).Objectives of this study are to (i) assess the water quality of lakes in Vrishabhavathi valley in Greater Bangalore, Karnataka;(ii) record the phytoplankton composition of lakes; (iii) evaluate the role of environmental variables in phytoplankton community structure using multivariate statistical analysis (canonical correspondence analysis, CCA) and (iv) biomonitoring of lakes by determining the Palmer algal genera pollution index.

Materials and Methods

Study Area: Select Lakes in Vrishabhavathi Valley (V Valley), Greater Bangalore

Greater Bangalore (12°49'5"N to 13°08'32"N latitude and 77°27'29"E to 77°47'2"E longitude), covers an area of 741 square kilometers. The district is located at an altitude of 920 m above mean sea level, delineating three watersheds, namely Koramangala and Challaghatta Valley (K and C Valley), Vrishabhavathi Valley (V Valley), and Hebbal Valley (H Valley). Due to the undulating terrain, a large number of tanks were created by the earlier rulers in Bangalore that provide water for various uses like drinking, irrigation, washing,fishing, *etc.* The mean annual rainfall in Bangalore is 880 mm, with 60 rainy days a year over a decade. The summer temperature ranges from 18°C - 38°C while the winter temperature ranges from 12°C - 25°C (Ramachandra and Aithal, 2015).Vrishabhavathi valley has a catchment area of ~170 square kilometers that covers 97 wards of Bengaluru. Vrishabhavathi river had numerous interconnected lake systems such as Kempambudilake in Basavanagudi, Sankey lake in Sadashivnagar, Yediyurkerein Yediyur, *etc.* (Ramachandra *et al.*, 2017). The present study involves field survey, physico-chemical analysis, and recording phytoplankton diversity of 15lakes in Vrishabhavathi Valley (V Valley) of Greater Bangalore (Figure 10.1).

Water Quality Analyses of Lakes

Water samples were collected from 15 lakes in Vrishabhavathi valley for physico-chemical analysis during 2017 - 2018. The physico-chemical parameters analyzed include water temperature (WT), total dissolved solids (TDS), electrical conductivity (EC), dissolved oxygen (DO), pH, chloride (Cl), total alkalinity (TA),total

Figure 10.1: Lakes of Vrishabhavathi Valley, Greater Bangalore, Karnataka, India.

hardness (TH), calcium (Ca), magnesium (Mg), biochemical oxygen demand (BOD), chemical oxygen demand (COD), nitrate (N) and ortho-phosphate (OP). The on site parameters include water temperature, pH, dissolved oxygen, total dissolved solids, and electrical conductivity measured using a portable meter (Extech). The turbidity was measured with a hand-held Turbidometer (Hach). The water samples collected from monitored lakes in clean and sterile polypropylene bottles were transported to the laboratory for further chemical and nutrient analysis using standard methods (APHA, 2005), and experiments were carried out in triplicates.

QGIS open-source GIS software was used for generating maps of various physico-chemical parameters of 15 lakes in the Vrishabhavathi valley of Greater Bangalore.

Phytoplankton Sampling and Analyses

Planktons were collected by filtering 50 liters of water from the lake surface using a plankton net. Phytoplankton samples were fixed with 4 per cent Lugol iodine solution and concentrated to 50 mL. Then, the samples were stored in labeled plastic bottles and transported to the laboratory for taxonomic analysis. Phytoplankton samples were identified according to the standard keys (Desikachary, 1959; Prescott, 1970; AlgaeBase, 2021)and enumerated with an optical microscope at 10 X and 40 X magnifications. Generally, three subsamples were counted for each lake sample. The maps depicting phytoplankton diversity of 15 lakes in Vrishabhavathi valley were created using QGIS open-source GIS software.

Data Analysis

Statistical analyses were carried out using PAST3 software. Pearson correlation analysis was carried out to evaluate the correlation between phytoplankton abundance and water quality parameters. Then, multivariate statistical analysis of the water quality data set and phytoplankton was performed through (CCA) to understand the relationship between water quality parameters and phytoplankton distribution as well as the composition of the monitored lakes. In this study, the Shannon-Wiener index, Simpson index, Pielou Evenness index, Berger and Parker index, Margalef Diversity index, and Menhinick index were used to evaluate the phytoplankton structure lakes. Various diversity measures such as the Shannon–Wiener index (H′), Evenness (J′), Simpson index (Ds), richness (S), and Berger and Parker index (Db) can be used to estimate the phytoplankton diversity (Zhang *et al.*, 2016).

Results and Discussion

Variations in Physico-chemical Parameters of Lakes

The lakes monitored in this study varied in their physico-chemical characteristics as these lakes receive untreated sewage or pollutants. An alteration of the physical, chemical, and biological properties of water happens due to pollution with the sustained inflow of untreated or partially treated wastewater from domestic and industrial sources (Udhayakumar *et al.*, 2016). The changes in water quality can alter phytoplankton community structure and distribution in lakes. The discussions on physico-chemical parameters analyzed are elaborated below:

Water Temperature and pH

Temperature strongly influences the chemical composition of cells, uptake of nutrients, uptake of carbon-dioxide, and growth rate of phytoplankton (Singh and Singh, 2015; Brönmark and Hansson, 2002). Water temperature varied between

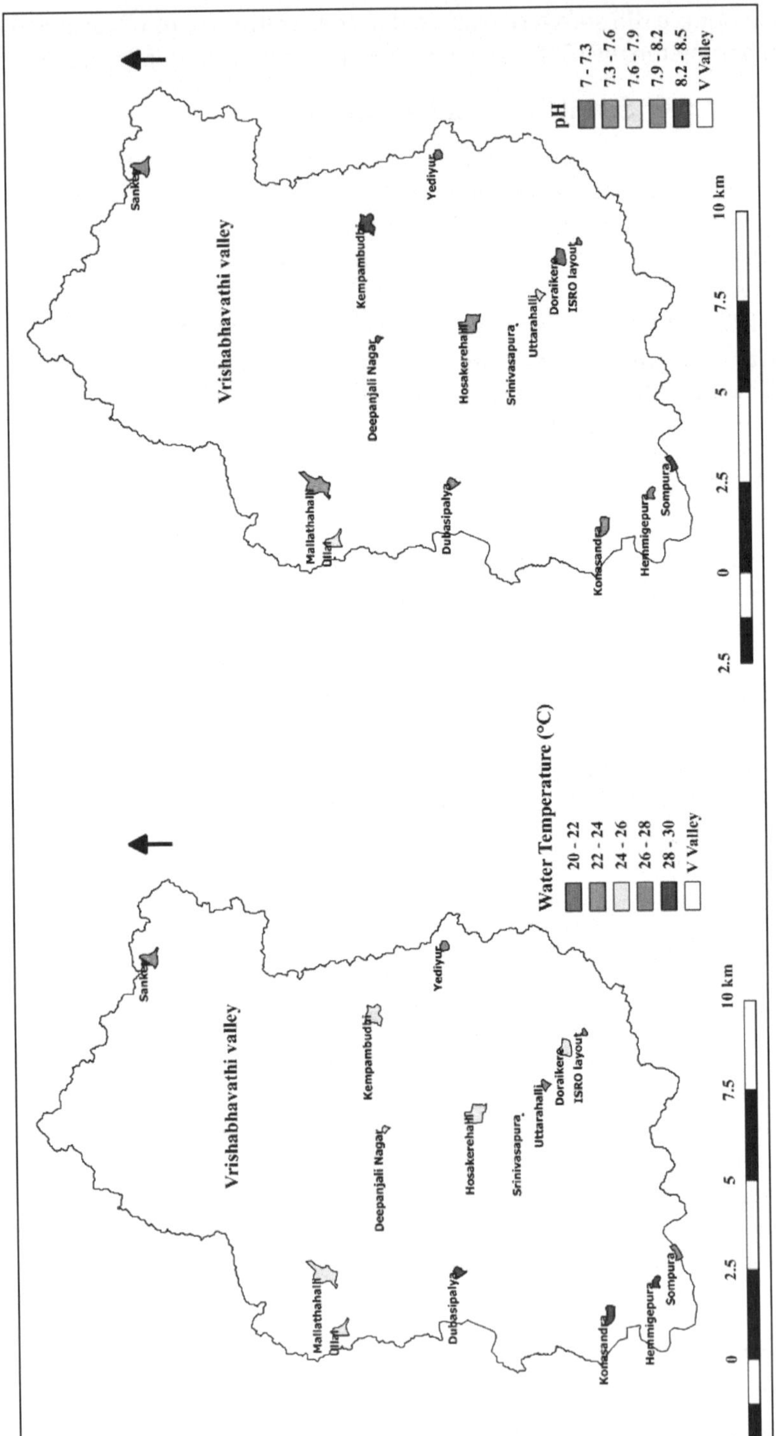

Figure 10.2: Variation of Water Temperature and pH in Lakes of Bangalore.

23.9 °C to 29.7 °C in monitored lakes during the current study (Figure 10.2). pH indicates hydrogen ion concentration in lake water (Kumar and Mahajan, 2020) and is generally governed by the equilibrium between carbon-dioxide, carbonate, and bicarbonate ions. Here, alkaline conditions prevailed in lakes, which ranged from 7.01 – 8.5. This range of pH is suitable for phytoplankton growth. During photosynthetic activity by phytoplankton, pH increases due to the consumption of carbon-dioxide.

Total Dissolved Solids (TDS) and Electrical Conductivity (EC)

The total dissolved solids (TDS) include inorganic salts and small amounts of organic matter in water. The concentration of total dissolved solids ranged from 152.5 mg/L to 883.42 mg/L in lakes (Figure 10.3). The variations in TDS occur naturally from rocks and soil as well as from anthropogenic sources through sewage, industrial effluent discharges, and urban run-off (Ntengwe, 2006). The conductivity is a measure of the capacity of water to conduct an electric current. The value of electrical conductivity ranged from 306.67 μS/cm to 1303.67 μS/cm (Figure 10.3). EC depends on the type, quantity, and ionic state of dissolved substances (electrolytes), thus, indicates the mineral content of water (Le *et al.*, 2017). Mallathahalli lake had the highest level of TDS and EC, hence shows ionic pollution.

Total Hardness, Calcium, and Magnesium

Hard water is not suitable for domestic and industrial purposes. The cations that impart total hardness to water are calcium, magnesium, iron, strontium, and manganese. The total hardness, calcium, and magnesium varied among sampled lakes and ranged from 78.67 mg/L- 463.5 mg/L, 23.51 mg/L - 117.9 mg/L, and 4.85 mg/L - 41.09 mg/L,respectively (Figure 10.4). Mallathahalli lake had the highest levels of total hardness, calcium, and magnesium, which indicates ionic pollution.

Total Alkalinity and Chloride

Total alkalinity is determined by the concentration of hydroxide, carbonate, and bicarbonate in water. Thus, total alkalinity influences the composition and abundance of phytoplankton (Bose *et al.*, 2019).

Mallathahalli lake receives a lot of untreated domestic and industrial sewage, so profuse growth of phytoplankton was observed that increased the alkalinity of lake water. In addition, the discharge of domestic and industrial wastes increases chloride levels in lake water. The concentration of total alkalinity and chloride in lakes ranged from 81.33 mg/L - 684.33mg/L and 26.74 mg/L - 305.89 mg/L, respectively (Figure 10.5).

Turbidity and Dissolved Oxygen (DO)

Turbidity in water occurs mainly due to particles of clay, silt, organic particles, inorganic matter, micro-organisms, and algae bloom. The range of turbidity in monitored lakes was found as 5.77 NTU - 103.51 NTU (Figure 10.6). Under

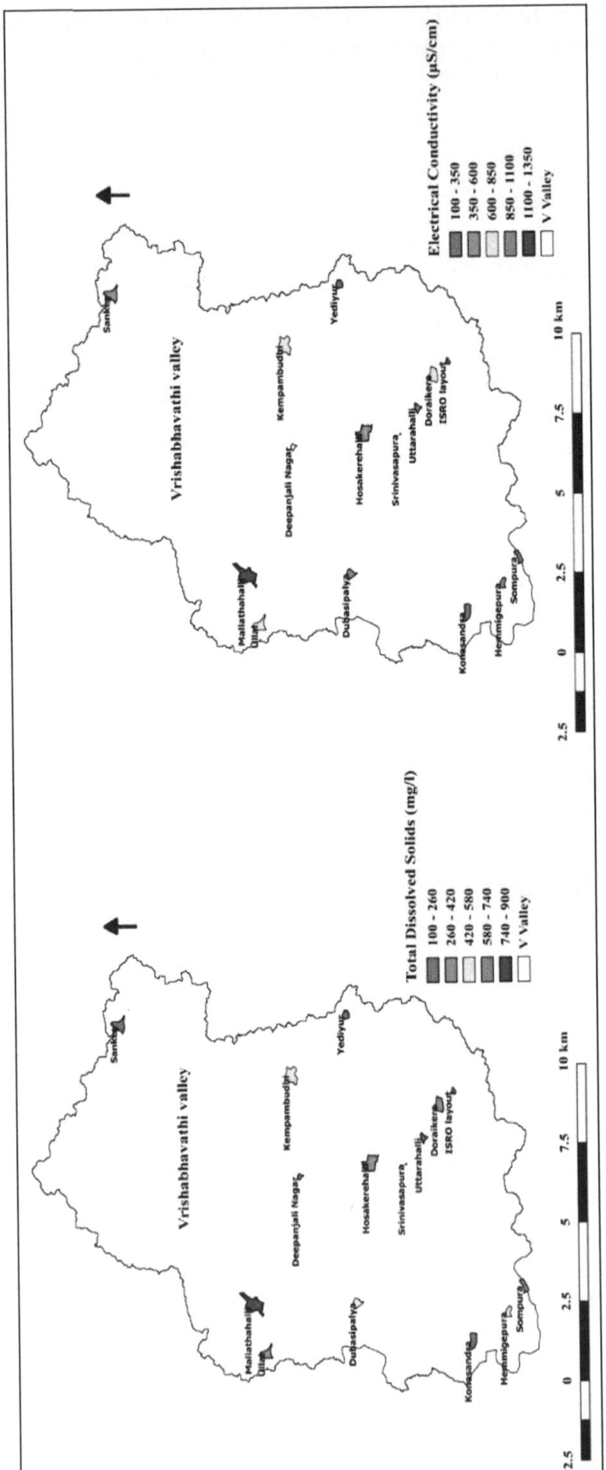

Figure 10.3: Level of Total Dissolved Solids and Electrical Conductivity in Lakes.

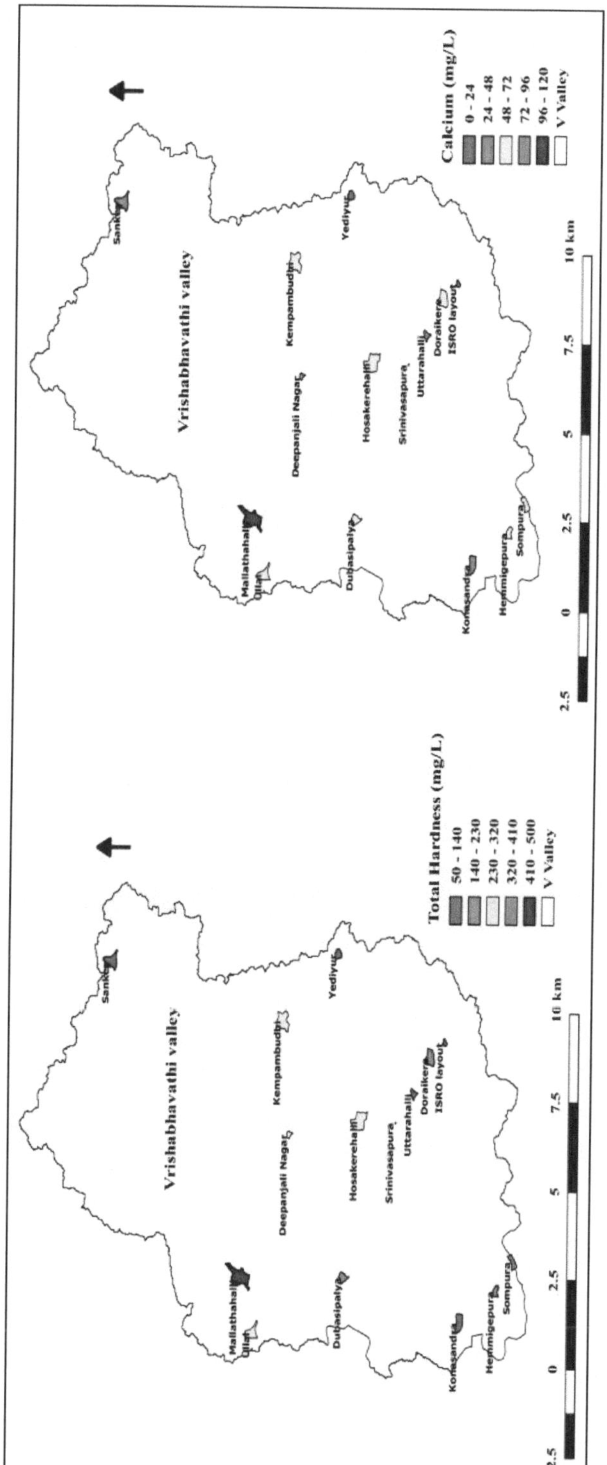

Figure 10.4: Concentration of Total Hardness and Calciumin Lakes.

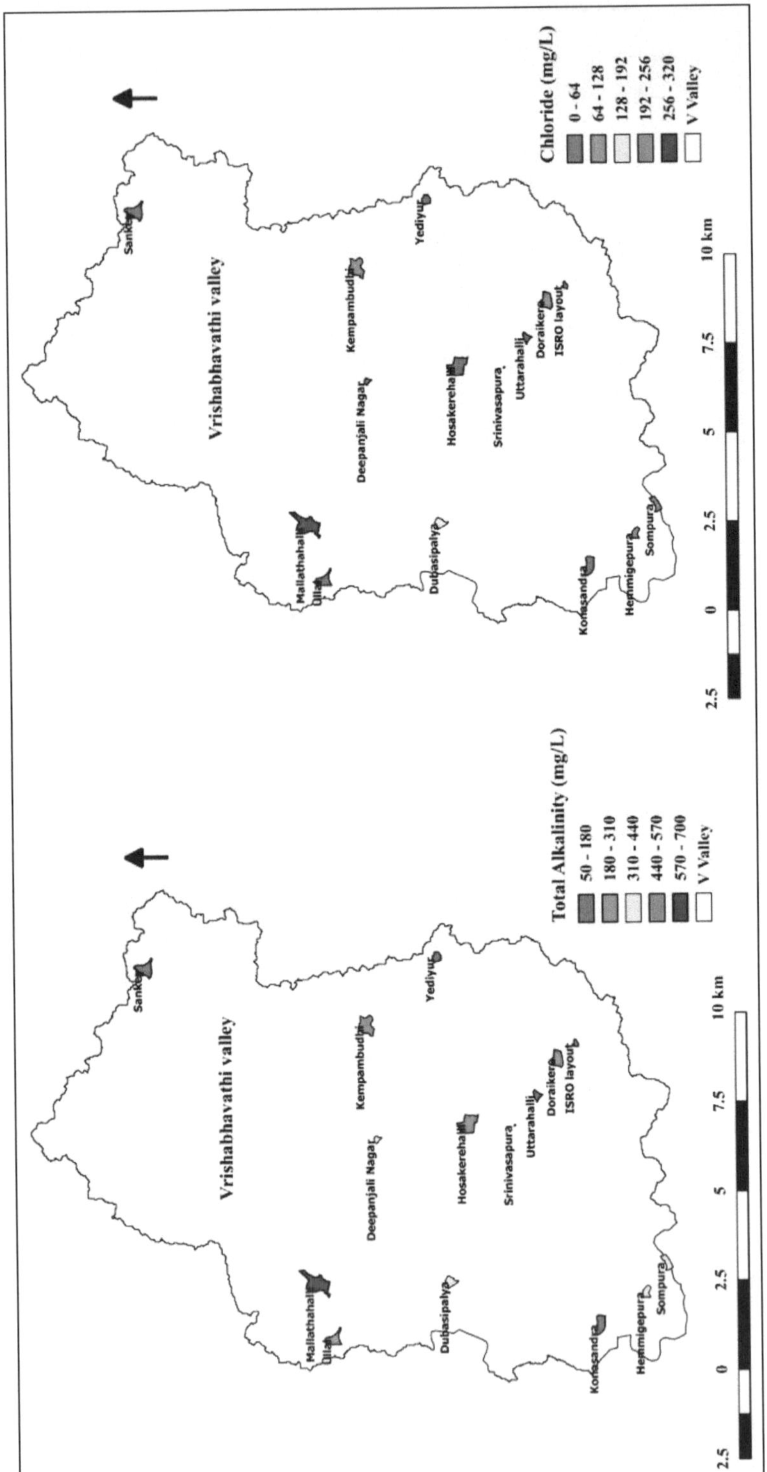

Figure 10.5: Concentration of Total Alkalinity and Chloride in Lakes.

turbid conditions, only less light can penetrate through the water column to the lake bottom. This results in inhibition of the growth of phytoplankton and lower photosynthetic activity (Yang *et al.*, 2012). This may, in turn, decrease the dissolved oxygen level in lake water. The dissolved oxygen levels in lakes ranged from 0.81 mg/L - 9.84 mg/L (Figure 10.6). Dissolved oxygen affects the various physico-chemical attributes and biological processes in lake water (Ramachandra *et al.*, 2014).

Biochemical Oxygen Demand (BOD) and Chemical Oxygen Demand (COD)

Biochemical oxygen demand (BOD) determines the amount of oxygen required by aerobic micro-organisms to stabilize the organic matter (Avvannavar and Shrihari, 2008). High COD concentrations indicate pollution related to untreated domestic sewage and industrial effluents. The BOD and COD of monitored lakes ranged between 12.2 mg/L - 60.97 mg/L and 21.33 mg/L - 128 mg/L, respectively (Figure 10.7). The higher levels of BOD and COD show the level of pollution, which will increase the oxygen demand in lakes. Among the sampled lakes, Kempambudhi lake had more elevated BOD and COD, indicating organic pollution.

Nitrate and Ortho-phosphate

Nutrients such as nitrogen and phosphorus are the limiting factors for phytoplankton growth. In the present study, nitrate and ortho-phosphate ranged between 0.01 mg/L - 3.1 mg/L and 0.077 mg/L - 3.701 mg/L, respectively (Figure 10.8). The excessive amount of nitrogen and phosphorus can stimulate eutrophic conditions (He *et al.*, 2011). The discharge of phosphorus into aquatic ecosystems results in eutrophication, whichhas enhanced the overgrowth of algae and macrophytes (Çelekli and Sahin, 2021). Kempambudhi lake was found to be rich in nitrate and ortho-phosphate (OP), which indicates nutrient pollution.

The study reveals that (i) Mallathahalli lake had higher levels of TDS, EC, total hardness, calcium, magnesium, chloride, total alkalinity, (ii) Kempambudhi lake with high pH, BOD, COD, and OP show organic and nutrient pollution, and (iii)Yediyur lake had less ionic contents such as TDS, total hardness, calcium, magnesium, and COD. Thus, the monitored lakes are polluted as they receive untreated domestic waste, agricultural residues, and industrial effluents. This may threaten biodiversity, reduce the aesthetic and recreational value of lakes.

Phytoplankton Community in Lakes

The present study covers 15 lakes in Vrishabhavathi valley of Bangalore, which recorded a total of 58 genera of phytoplankton belonging to 5 phyla, 9 class, 24 orders, and 41 families were identified (Figure 10.9). The fifty-eight species of phytoplankton were represented by five major groups, namely, Chlorophyta (25 species), Bacillariophyta (12 species), Cyanobacteria (12 species), Charophyta (5 species), and Euglenozoa (4 species).

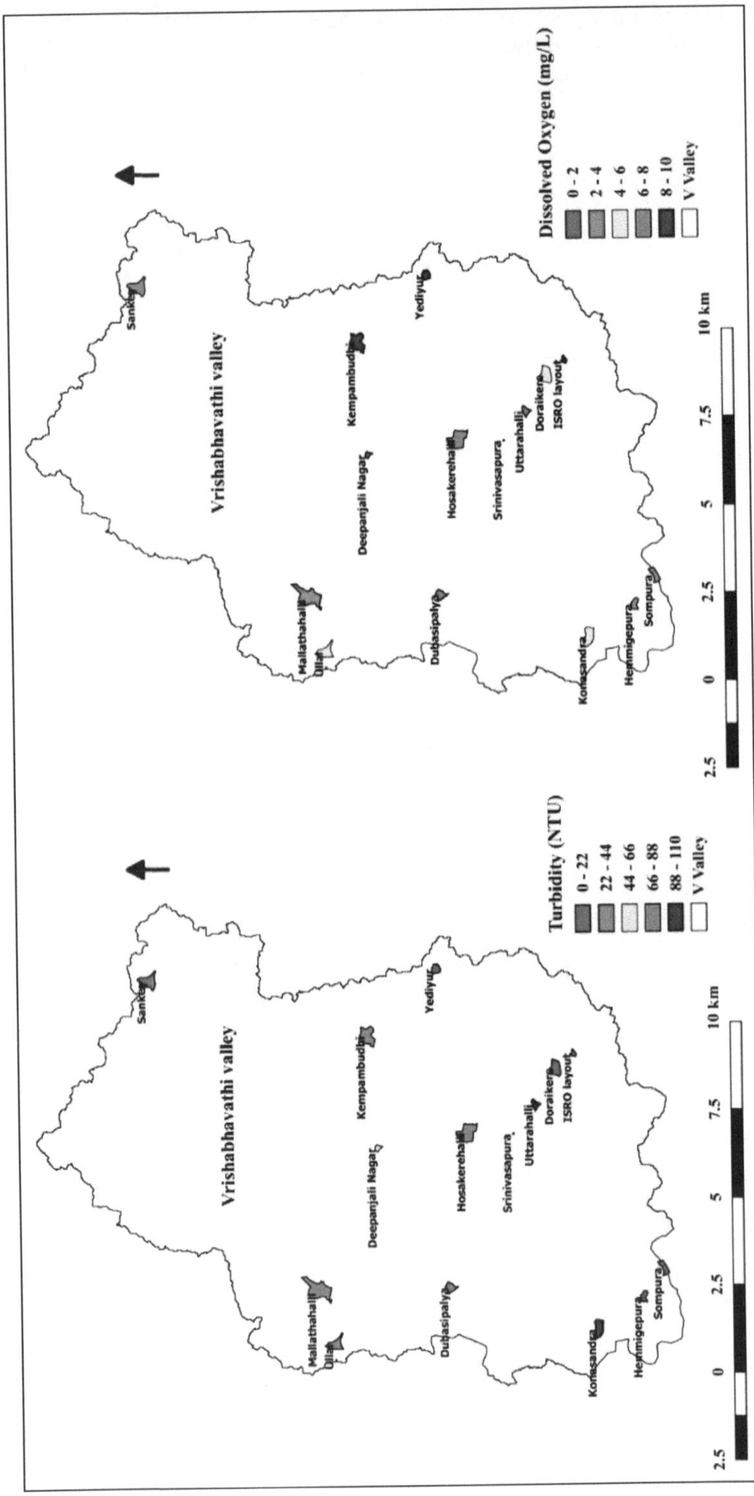

Figure 10.6: Turbidity and Dissolved Oxygen Levels in Lakes.

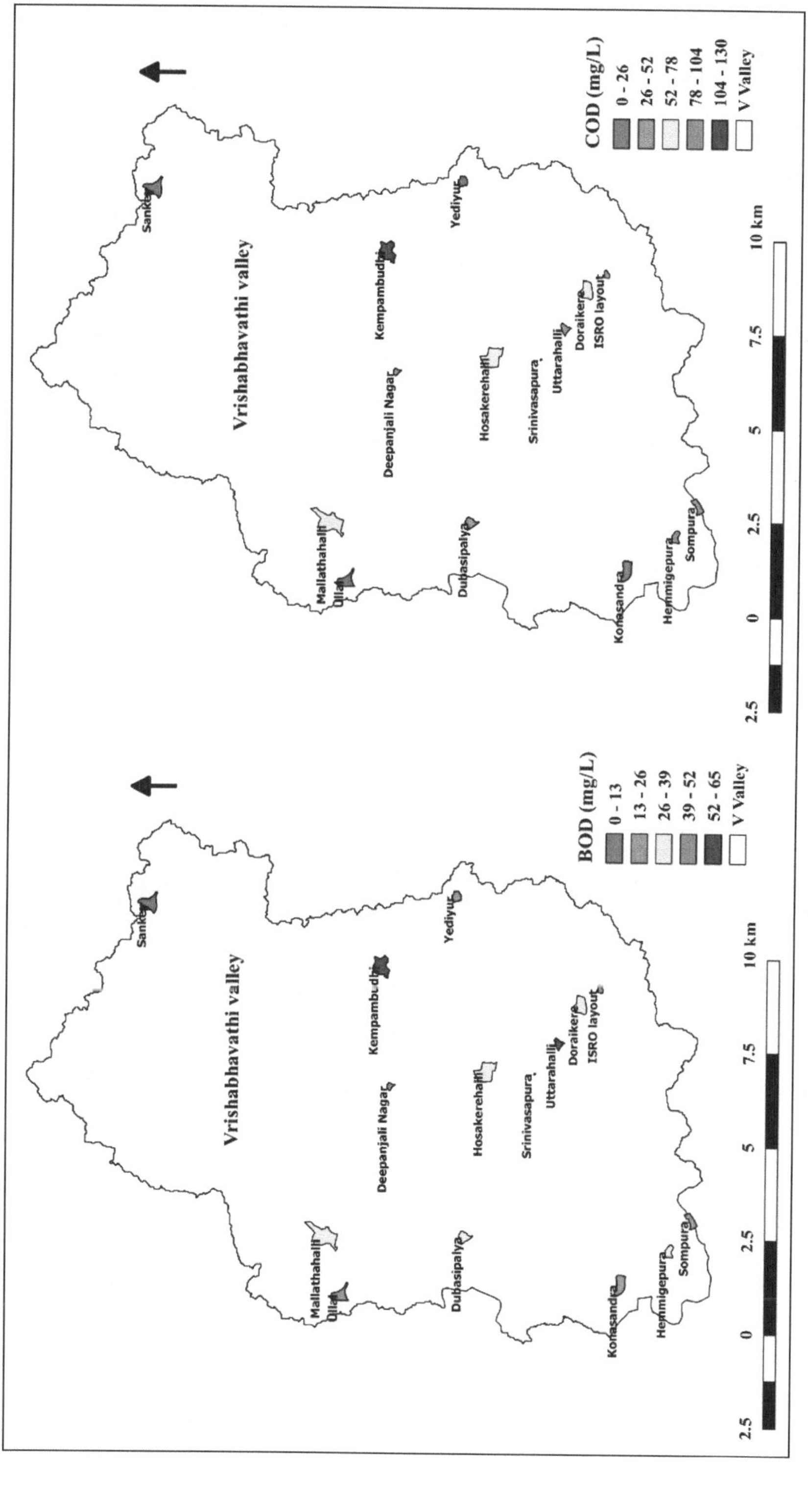

Figure 10.7: Variation of BOD and COD in Lakes.

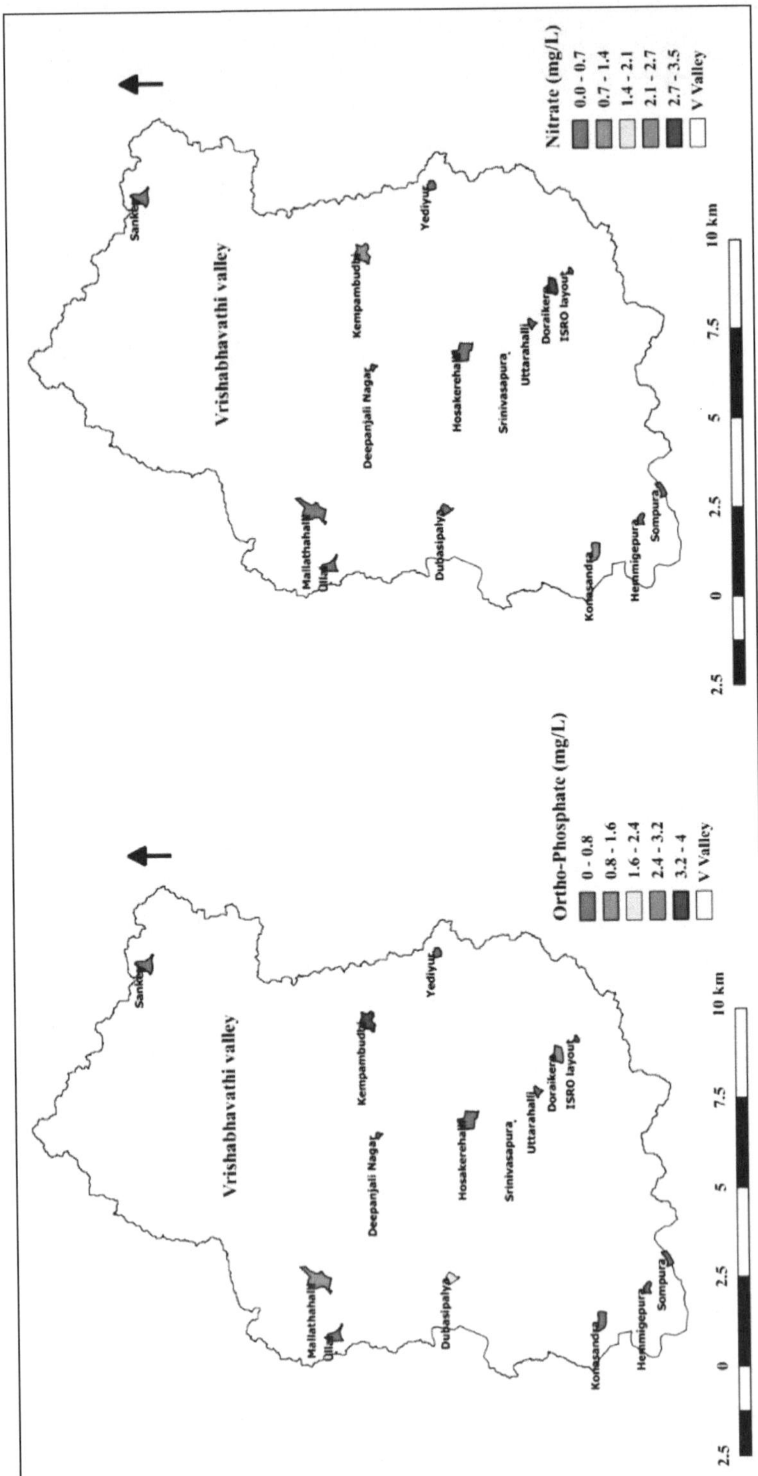

Figure 10.8: Concentration of Nitrate and Ortho-phosphate in Lakes.

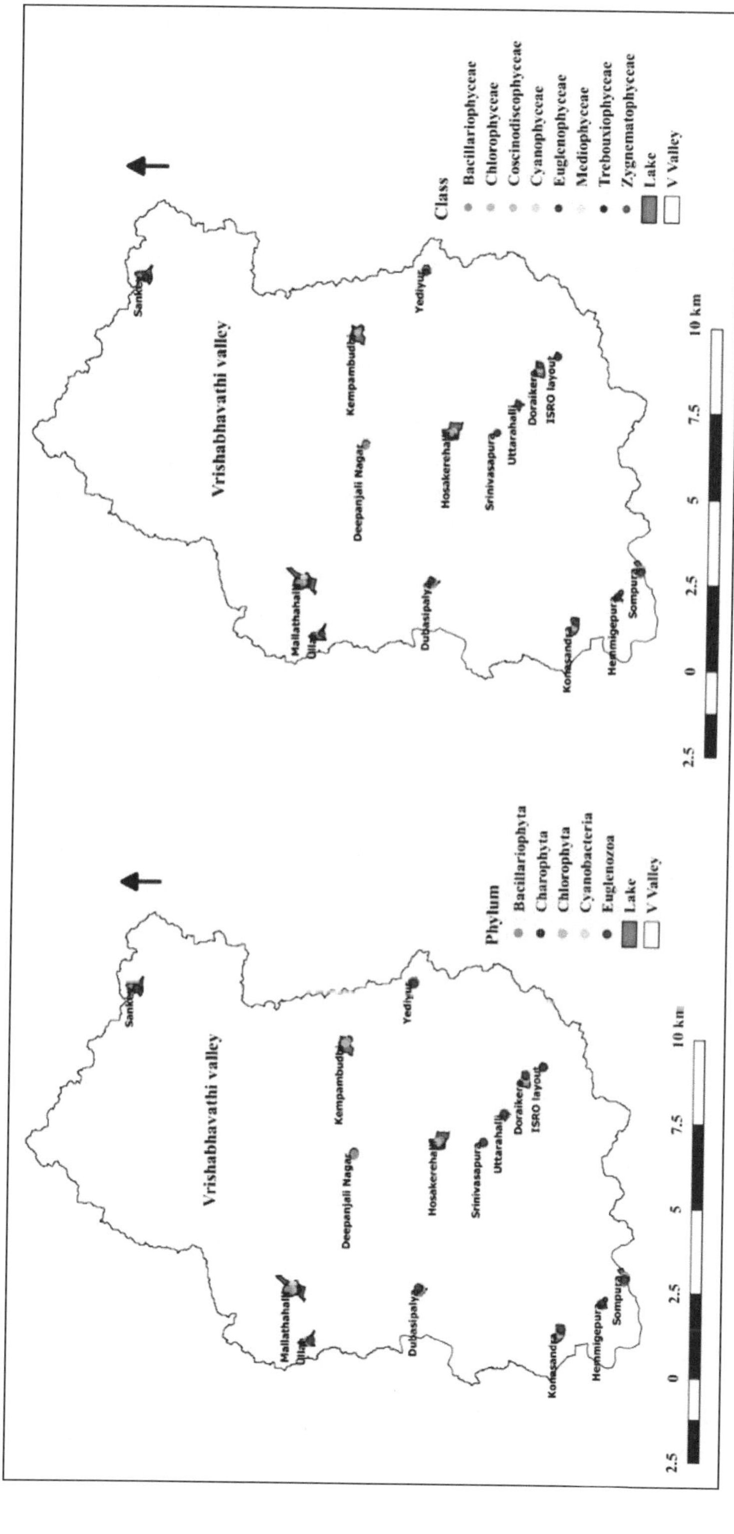

Figure 10.9: Phytoplankton Composition (Phylum and Class) in Lakes of Bangalore.

Cyanobacteria, commonly known as blue-green algae, are photosynthetic prokaryotes that form algal blooms in water (Elliott, 2012). Cyanobacteria are represented by species such as *Anabaena* sp., *Aphanocapsa* sp., *Chroococcus* sp., *Coelosphaerium* sp., *Cylindrospermopsis* sp., *Gloeocapsa* sp., *Merismopedia* sp., *Microcystis* sp., *Oscillatoria* sp., *Phormidium* sp., *Planktothrix* sp. and *Spirulina* sp. Cyanobacteria, mainly *Microcystis* sp.was dominant and formed surface blooms in Doraikere, Uttarahalli, ISRO layout, and Yediyur lakes in Vrishabhavathi valley, Greater Bangalore, Karnataka (Figure 10.10). A total of 25 species of Chlorophyta was recorded in this study, such as *Actinastrum* sp., *Ankistrodesmus* sp., *Asterococcus* sp., *Botryococcus* sp., *Chlamydomonas* sp., *Chlorella* sp., *Chlorococcum* sp., *Coelastrum* sp., *Crucigenia* sp., *Desmodesmus* sp., *Dictyosphaerium* sp., *Golenkinia* sp., *Kirchneriella* sp., *Micractinium* sp., *Monoraphidium* sp., *Oocystis* sp., *Pandorina* sp., *Pediastrum* sp., *Scenedesmus* spp., *Schroederia* sp., *Stichococcus* sp., *Stigeoclonium* sp., *Tetraedron* sp. and *Tetrastrum* sp. Both pennate and centric forms of Bacillariophyta (or diatoms) are present lakes inVrishabhavathi valley, Bangalore.*Achnanthes* sp., *Amphora* sp., *Cyclotella* sp., *Fragilaria* sp., *Gomphonema* sp., *Melosira* sp., *Navicula* sp., *Nitzschia* sp., *Pinnularia* sp., *Stauroneis* sp., *Surirella* sp. and *Synedra* sp. belonging to Bacillariophyta was recorded.

A total of four species under Charophyta were recorded, such as *Closterium* sp., *Cosmarium* sp., *Euastrum* sp., *Spirogyra* sp. and *Staurastrum* sp. in the present investigation. Euglenozoa is represented by species such as *Euglena* spp., *Lepocinclis* sp., *Phacus* sp. and *Trachelomonas* sp. *Euglena* sp. is mixotrophic algae, which can grow faster in wastewater because they can utilize organic matter as an energy source (Mahapatra *et al.,* 2013).

Diversity Indices

Various diversity measures were used to estimate the phytoplankton diversity of lakes in Vrishabhavathi valley, Bangalore. Shannon-Wiener index ranged from 0.16 -2.54. Shannon and Wiener index value of 0.0-1.0 indicates heavy pollution, 1.0-2.0 indicates moderate pollution, 2.0-3.0 indicates light pollution, and 3.0-4.5 indicates slight pollution. The species diversity indices decrease in polluted waters (Shanthala *et al.,* 2009). Thus, Doraikere (2.54) is heavily polluted with less species diversity. Simpson index ranged from 0.045 (Doraikere) -0.882 (Sompura, Mallathahalli). Pielou evenness index reveals an even distribution of individuals among varied species (Miao *et al.,* 2019). The evenness index ranges from near 0 (low evenness) to 1 (dominance). Evenness index ranged from 0.07 (Uttarahalli) – 0.80 (Deepanjali Nagar, Srinivasapura). Menhinick index ranged from 0.26 (Doraikere) – 1.59 (Hemmigepura) in this study. Margalef species richness index measures the number of different species in each sample (Ray *et al.,* 2021). Margalef index ranged from 0.72 – 4.77 as Dubasipalya lake has the highest taxa whereas Srinivasapura has the lowest taxa and individuals. A high value of richness index indicates healthy status, while a low value indicates the unhealthy and contaminated status (Hussien and Zaghloul, 2017). The Berger-Parker index gives the measure

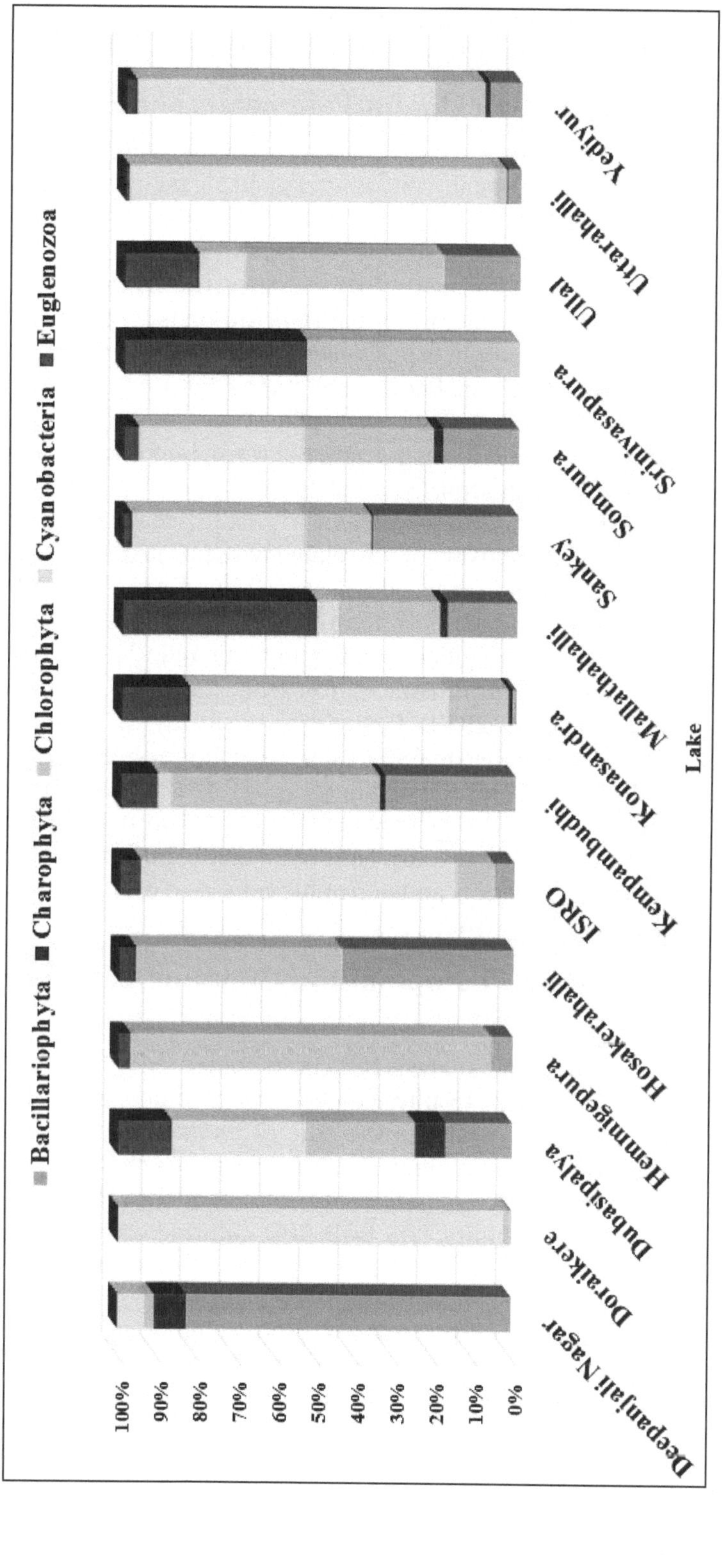

Figure 10.10: Phytoplankton Composition in Lakes of Bangalore.

of the most abundant species (Berger and Parker, 1970). Berger-Parker index in this study ranged from 0.21 (Deepanjali Nagar) -0.98 (Doraikere).

Correlation between Water Quality Parameters and Phytoplankton

Pearson correlation analysis is used to measure the interrelation and extent of associations among the water quality parameters and algal phyla. Bacillariophyta negatively correlated with BOD (r = -0.31) and COD (r = -0.38). Bacillariophyta responds rapidly to organic and nutrient contamination (Saxena *et al.,* 2020). Charophyta positively correlated with water temperature (r = 0.41), electrical conductivity (r = 0.36), magnesium (r = 0.40), ortho-phosphate (r = 0.31) and Bacillariophyta (r = 0.57). Chlorophyta positively correlated with water temperature (r = 0.38), Charophyta (r = 0.68) and Bacillariophyta (r = 0.90). A positive correlation was found between temperature and phytoplankton such as Charophyta and Chlorophyta (Gogoi *et al.,* 2019) as temperature plays an important role in growth and reproduction of phytoplankton. Cyanobacteria negatively correlated with pH (r = -0.45), total hardness (r = -0.44), calcium (r = -0.39), magnesium (r= -0.46), total alkalinity (r = -0.49) and positively correlated with nitrate (r = 0.63). A similar pattern of positive correlation recorded between Cyanophytes with nitrate, and inverse correlation with alkalinity, pH, total hardness, calcium and magnesium (Mishra *et al.,* 2019). In Baiyangdian lake, Cyanophyta and Chlorophyta became dominant because of increased organic matter from industrial wastewater and domestic sewage (Wang *et al.,* 2013). Euglenozoa positively correlated with TDS (r = 0.49), EC (r = 0.59), total hardness (r = 0.49), calcium (r = 0.36), magnesium (r = 0.61), chloride (r = 0.58), total alkalinity (r = 0.43), ortho-phosphate (r = 0.31), Bacillariophyta (r = 0.39), Charophyta (r = 0.79), Chlorophyta (r = 0.72) and negatively correlated with DO (r = -0.30). The dominance of Euglenophyta indicates organic pollution, water pollution, and eutrophic condition of water body (Shams and Shamsabadi, 2019).

Relationship between Phytoplankton Community and Physico-chemical Factors

Canonical correspondence analysis (CCA) was computed with phytoplankton composition data (58 genera), 15 physico-chemical parameters across 15 lakes during the study period to evaluate the role of environmental variables in structuring phytoplankton communities. For CCA, 15 physico-chemical parameters, namely: water temperature, pH, conductivity, total alkalinity, dissolved oxygen, nitrate, ortho-phosphate, turbidity, TDS, BOD, COD, chloride, calcium, magnesium, and total hardness,were selected. Different species of phytoplankton showed varied responses to physico-chemical parameters. The length of the arrow in CCA ordination indicates the relative importance of environmental variables. Eigenvalues of the first two axes are 0.58 and 0.48,which explain 34.97 per cent of the variance of phytoplankton and environmental data. Axis 1 has strong positive loadings of turbidity. Thus, *Spirulina* sp., *Actinastrum* sp., *Asterococcus* sp., *Micractinium* sp. and *Phacus* sp. are positively correlated with the turbidity. Turbidity of freshwater lakes increases due to the influx of urban run-off, silt, sand, and organic matter (Nandigam *et al.,* 2016). This decreases the phytoplankton diversity and increases the dominance of *Spirulina* sp. as in Konasandra lake.

Axis 2 has strong positive loadings of total hardness, calcium, magnesium, and total alkalinity. Total hardness, calcium, magnesium, and total alkalinity are found in lakes such as Deepanjali Nagar, Hemmigepura, Sompura, Kempambudhi, Srinivasapura, Ullal, Doraikere, Dubasipalya, Hosakerehalli, and Uttarahalli. Therefore, the growth of species such as *Anabaena* sp., *Chroococcus* sp., *Cymbella* sp., *Fragilaria* sp., *Gomphonema* sp., *Lepocinclis* sp., *Monoraphidium* sp., *Navicula* sp., *Oscillatoria* sp., *Pandorina* sp., *Phormidium* sp., *Pinnularia* sp., *Spirogyra* sp., *Stauroneis* sp., *Stichococcus* sp. and *Surirella* sp. are positively dependent on the parameters like total hardness, calcium, magnesium, and total alkalinity. Magnesium is crucial for chlorophyll synthesis and serves as a limiting factor for phytoplankton growth and reproduction (Narchonai *et al.,* 2019). The remaining species did not show a relation with the selected variables. Phytoplankton species such as *Amphora* sp., *Aphanocapsa* sp., *Botryococcus* sp., *Chlorococcum* sp., *Desmodesmus* sp., *Dictyosphaerium* sp., *Euastrum* sp., *Golenkinia* sp., *Kirchneriella* sp., *Melosira* sp., *Microcystis* sp., *Pediastrum* sp., *Planktothrix* sp., *Scenedesmus* spp., *Schroederia* sp., *Staurastrum* sp., *Stigeoclonium* sp., *Tetraedron* sp., *Tetrastrum* sp. and *Trachelomonas* sp. found nearer to the center of CCA ordination has a broad range of tolerance to the variations in physico-chemical parameters in lakes. *Anabaena* sp., *Merismopedia* sp., *Microcystis* sp. and *Oscillatoria* sp. indicate eutrophic lake conditions (Yilmaz *et al.,* 2018).

Palmer Index

The phytoplankton community composition alters quickly due to changes in the physico-chemical characteristics of water. For example, the presence of varied species of phytoplankton can indicate the quality of water. For example, Oscillatoria and Euglena are indicators of dirty water, while Pediastrum indicates clean water (Kantachote *et al.,* 2009). Thus,phytoplankton communities serve as indicators of pollution in an aquatic ecosystem. The Palmer index is a rapid, efficient, and cost-effective tool that reveals the water quality status or water pollution in freshwater bodies (Salem *et al.,* 2017). Palmer prepared a list of algal genera and species that are tolerant to varying levels of organic pollution (Vishal and Meeta, 2020). Palmer index accounts for the abundant pollution tolerant phytoplankton taxa and categorizes water bodies based on the pollution status. The palmer index value of 20 or more confirms high organic pollution. A value of 15-20 indicates probable organic pollution. The values between 10-15 and 0-10 show moderate pollution and lack of organic pollution, respectively (Jose and Kumar, 2011).

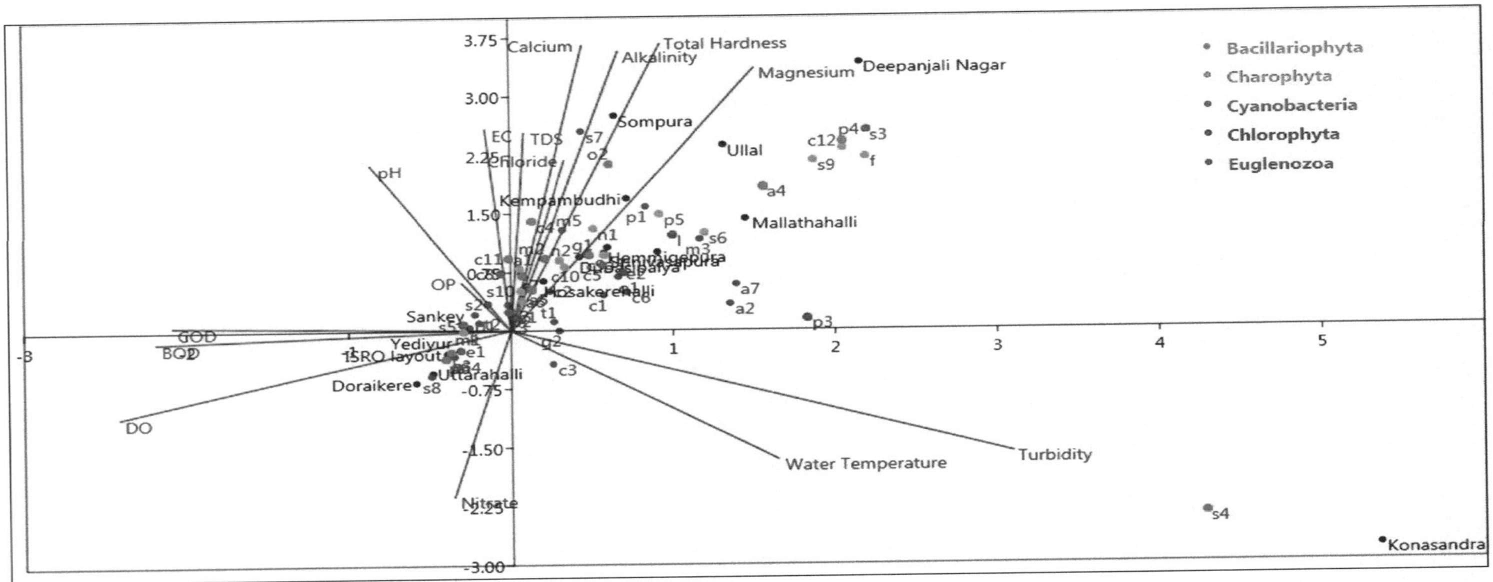

Figure 10.11: Canonical Correspondence Analysis (CCA) Ordination Diagram Showing Scatter Plot for Phytoplankton, Lakes and Physico-chemical Parameters (Arrows)

(Abbreviations: a1: *Achnanthes* sp.; a2: *Actinastrum* sp.; a3: *Amphora* sp.; a4: *Anabaena* sp.; a5: *Ankistrodesmus* sp.; a6: *Aphanocapsa* sp.; a7: *Asterococcus* sp.; b: *Botryococcus* sp.; c1: *Chlamydomonas* sp.; c2: *Chlorella* sp.; c3: *Chlorococcum* sp.; c4: *Chroococcus* sp.; c5: *Closterium* sp.; c6: *Coelastrum* sp.; c7: *Coelosphaerium* sp.; c8: *Cosmarium* sp.; c9: *Crucigenia* sp.; c10: *Cyclotella* sp.; c11: *Cylindrospermopsis* sp.; c12: *Cymbella* sp.; d1: *Desmodesmus* sp.; d2: *Dictyosphaerium* sp.; e1: *Euastrum* sp.; e2: *Euglena* spp.; f- *Fragilaria* sp.; g1: *Gloeocapsa* sp.; g2: *Golenkinia* sp.; g3: *Gomphonema* sp.; k: *Kirchneriella* sp.; l: *Lepocinclis* sp.; m1: *Melosira* sp.; m2: *Merismopedia* sp.; m3: *Micractinium* sp.; m4: *Microcystis* sp.; m5: *Monoraphidium* sp.; n1: *Navicula* sp.; n2: *Nitzschia* sp.; o1: *Oocystis* sp.; o2: *Oscillatoria* sp.;p1: *Pandorina* sp.; p2: *Pediastrum* sp.; p3: *Phacus* sp.; p4: *Phormidium* sp.; p5: *Pinnularia* sp.; p6: *Planktothrix* sp.; s1: *Scenedesmus* spp.; s2: *Schroederia* sp.; s3: *Spirogyra* sp.; s4: *Spirulina* sp.; s5: *Staurastrum* sp.; s6: *Stauroneis* sp.; s7: *Stichococcus*sp.; s8: *Stigeoclonium* sp.; s9: *Surirella* sp.; s10- *Synedra* sp.; t1: *Tetraedron* sp.; t2: *Tetrastrum* sp. and t3: *Trachelomonas* sp.).

According to Palmer index (Table 10.1), Doraikere, Deepanjali Nagar, Srinivasapura, Hosakerahalli lakes had moderate pollution as the Palmer index for these lakes was 10 -15.On the other hand, lakes such as ISRO Layout, Uttarahalli, Sankey, Yediyur, Sompura, Ullal, Hemmigepura, Konasandra, Dubasipalya, Mallathahalli, and Kempambudhi had high organic pollution (with Palmer index of 20 or more) in the present study. The dominance of species such as Chlorella, Oscillatoria, Pediastrum, Scenedesmus, Nitzschia, Melosira, Gomphonema, Navicula, Euglena, *etc.*, indicates of organic pollution in water bodies (Kshirsagar, 2013; Parvez *et al.*, 2019). The Palmer algal genera pollution index was in the range of 15 -24 in Vaigai river in Madurai (Noel and Rajan, 2015), while 5 - 28.6 in Chulband River in Maharashtra,which indicate organic pollution (Shahare, 2017). Palmer pollution index revealed that Sidi Abderrahmane is an oligotrophic or mesotrophic reservoir (Belokda *et al.*, 2019). Any alteration in the water quality of the aquatic ecosystem will alter the phytoplankton community structure. Only pollution tolerant species can thrive well in polluted waters, whereas sensitive species disappear, as they cannot survive in polluted water. Thus, phytoplankton serves as a bioindicator for assessing the extent of pollution of water bodies (Abdulwahid, 2016).

Table 10.1: Palmer Pollution Index of Lakes Based on algal genus

Lake	*Index Level*	*Remark*
ISRO	21	High organic pollution
Uttarahalli	27	High organic pollution
Sankey	25	High organic pollution
Yediyur	20	High organic pollution
Doraikere	14	Moderate pollution
Sompura	35	High organic pollution
Ullal	35	High organic pollution
Hemmigepura	26	High organic pollution
Konasandra	30	High organic pollution
Dubasipalya	39	High organic pollution
Mallathahalli	23	High organic pollution
Deepanjali Nagar	10	Moderate pollution
Srinivasapura	10	Moderate pollution
Hosakerehalli	12	Moderate pollution
Kempambudhi	25	High organic pollution

The physico-chemical parameters are the major factors controlling the structure and composition of phytoplankton in freshwater ecosystems. Phytoplanktons are an important component of an aquatic ecosystem and are responsible for primary productivity. The development of phytoplankton was favored by higher concentrations of nutrients and organic matter (Devercelli and

O'Farrell, 2013). Water quality parameters such as temperature, pH, DO, EC, BOD, COD, phosphate, and nitrate influenced phytoplankton species composition and distribution (Cheraghpour *et al.,* 2013). Phytoplankton grows well in nutrient-rich water and removes BOD, COD, and nutrients while providing oxygen to other aquatic organisms (Kiran *et al.,* 2016). The abundance of *Chlamydomonas* sp., *Cyclotella* sp. and Euglenophyta indicates the excessive availability of N and P as these species are organic-tolerant (Ghobara and Salem, 2017). Domestic sewage rich in sodium triphosphate (a component in synthetic detergents) can stimulate the growth and proliferation of *Chlorella vulgaris* (Marchello *et al.,* 2015). *Cyclotella meneghiniana*is an indicator of organic pollution (Krupa *et al.,* 2016). *Euglena* sp. is a biological indicator of high organic pollution (Mahapatra *et al.,* 2011). A shift in phytoplankton composition towards bloom-forming cyanobacteria occurs due to excess phosphorus loading (Lv *et al.,* 2011).

Cyanobacteria were found to be dominant in lakes of Vrishabhavathi valley, Bangalore. *Microcystis* sp. belonging to Cyanobacteria formed surface blooms in Doraikere, Uttarahalli, ISRO layout, and Yediyur lakes. Microcystis can tolerate high temperatures and hence survive well in warm, shallow, and eutrophic environments (Ke *et al.,* 2008). They form blooms in water bodies due to the buoyant nature imparted by the gas vacuoles (Muthukumar *et al.,* 2007). The surface blooms of *Microcystis* sp. will decrease the water transparency as well as light irradiation to bottom layers, which in turn inhibits the growth of benthic algae (Su *et al.,* 2014; Teta *et al.,* 2017). Harmful algal blooms (HABs) forms surface scum, induce hypoxia, alter the food web, and produce toxins thus, threatens ecosystem and human health (Fang *et al.,* 2019; Lu *et al.,* 2013; Catherine *et al.,* 2013). Excessive phytoplankton growth eventually causes depletion of dissolved oxygen (DO) and kills fish as well as other aquatic life. Cyanobacteria produce toxic secondary metabolites called cyanotoxins, such as hepatotoxins, dermatotoxins, neurotoxins, and cytotoxins (Du *et al.,* 2019; Beaver *et al.,* 2018). These cyanotoxins differ in their chemical structure and toxicology (Merel *et al.,* 2013). The hepatotoxic microcystins (MCs) produced by *Microcystis* sp. blooms at higher water temperature, pH, and dissolved oxygen (Wu *et al.,* 2014). Generally, species under Cyanophyceae are resistant to inorganic pollution, organic pollution, untreated effluents, and anaerobiosis (Marchello *et al.,* 2015). In Dongping lake, temperature and chemical oxygen demand (COD) are the main drivers of the cyanobacterial community composition (Lu *et al.,* 2013). Cyanobacteria biomass increased with a rise in total phosphorus concentration (Yang *et al.,* 2016). *Microcystis aeruginosa, Spirulina* sp., and *Oscillatoria* sp. are bioindicators of eutrophic conditions (Wijeyaratne and Nanayakkara, 2020). The cyanobacterial dominance was influenced by rainfall, flow, water temperature, EC, DO, pH, total nitrogen, nitrate, ammonia, total phosphorus, ortho-phosphate, chlorophyll–a, BOD, COD, total organic carbon, iron, and silicon dioxide content (Kim *et al.,* 2019). Cyanophyta grows faster under conditions such as increased temperature, salinity, and light intensity with less water turbulence (Kouhanestani *et al.,* 2019).

Water quality monitoring and management of lakes at regular time intervals are compulsory to control water pollution and deterioration of freshwater ecosystems. The integration of wastewater treatment plants with constructed wetlands and algal pond helps in efficient nutrient removal from urban lakes (Ramachandra *et al.*, 2018). The periodic monitoring of lakes will also aid in bloom management. It is necessary to control the levels of nutrients, dissolved oxygen, and organic matter content of water bodies to prevent algal blooms (Chen *et al.*, 2018). Aeration of water bodies will increase the dissolved oxygen levels and reduce bloom formation tendency (Ramachandra *et al.*, 2015). A ban on the use of phosphate-containing detergents and improved wastewater treatment can reduce the phosphorus (P) loads in lakes. Reductions in nutrient (N and P) input are requisite for effective long-term control of cyanobacterial algal blooms (Paerl *et al.*, 2011). The increasing water pollution and harmful cyanobacterial algal blooms due to nutrient enrichment in lakes pose a severe threat to freshwater resources and biodiversity.

Conclusion

The water quality parameters play a decisive role in the structure and composition of phytoplankton in aquatic ecosystems. The water quality and phytoplankton community structure varied among the monitored 15 lakes of Vrishabhavathi valley, Greater Bangalore, Karnataka. Mallathahalli lake had higher levels of ionic pollution, while Kempambudhi lake showed higher levels of organic and nutrient pollution. A total of 58 genera of phytoplankton belonging to 5 phyla such as Chlorophyta, Bacillariophyta, Cyanobacteria, Charophyta, and Euglenozoa were recorded in the monitored 15 lakes. Cyanobacteria formed blooms in Doraikere Uttarahalli, ISRO layout, and Yediyur lakes in Vrishabhavathi valley, threatening aquatic life. CCA results revealed that turbidity, total hardness, calcium, magnesium, and alkalinity influences phytoplankton community structure and composition. Palmer index revealed that Doraikere, Deepanjali Nagar, Srinivasapura, and Hosakerehalli lakes had moderate pollution. In contrast, ISRO, Uttarahalli, Sankey, Yediyur, Sompura, Ullal, Hemmigepura, Konasandra, Dubasipalya, Mallathahalli, and Kempambudhi lakes had high organic pollution. Hence, phytoplanktons are helpful as bioindicators of water quality and aid asa low-cost tool for monitoring lakes. This study will serve as a baseline for wetland protection and biodiversity conservation and help various stakeholders decide to control pollution from domestic and industrial sources.

Acknowledgments

We thank the ENVIS Division, The Ministry of Environment, Forests and Climate Change, Government of India, and Indian Institute of Science (IISc) for supporting to ecological research.

REFERENCES

Abd El-Karim, M.S. (2014). Epipelic algal distribution in Ismailia Canal and the possible use of diatoms as bioindicators and a biomonitoring tool. *The Egyptian Journal of Aquatic Research, 40*(4), 385-393. http: //dx.doi.org/10.1016/j.ejar.2014.12.005

Abdulwahid, K.D. (2016). Assessment of water quality in Khreisan river from Baquba city, Diyala (Iraq) by using algae as a bioindicator, *International Journal of Current Research, 8*(6), 33108-33115.

Adamczuk, M., Pawlik-Skowrońska, B., and Solis, M. (2020). Do anthropogenic hydrological alterations in shallow lakes affect the dynamics of plankton? *Ecological Indicators, 114,* 1-14. https: //doi.org/10.1016/j.ecolind.2020.106312.

Afonina, E.Y., and Tashlykova, N.A. (2018). Plankton community and the relationship with the environment in saline lakes of Onon-Torey plain, Northeastern Mongolia. *Saudi Journal of Biological Sciences, 25*(2), 399-408.

AlgaeBase, (2021). https: //www.algaebase.org/

Alves, J.D.P.H., Fonseca, L.C., Chielle, R.D.S.A., and Macedo, L.C.B. (2018). Monitoring water quality of the Sergipe river basin: An evaluation using multivariate data analysis. *RBRH, 23,* 1-12. https: //doi.org/10.1590/2318-0331.231820170124

APHA, AWWA, WEF. (2005). Standard methods for the examination of water and wastewater. 21st Edition. D.C., Washington: American Public Health Association.

Avvannavar, S.M., and Shrihari, S. (2008). Evaluation of water quality index for drinking purposes for river Netravathi, Mangalore, South India. *Environmental Monitoring and Assessment, 143*(1-3), 279-290. DOI 10.1007/s10661-007-9977-7

Beaver, J.R., Tausz, C.E., Scotese, K.C., Pollard, A.I., and Mitchell, R.M. (2018). Environmental factors influencing the quantitative distribution of microcystin and common potentially toxigenic cyanobacteria in US lakes and reservoirs. *Harmful Algae, 78,* 118-128. doi: 10.1016/j.hal.2018.08.004.

Belokda, W., Khalil, K., Loudiki, M., Aziz, F., and Elkalay, K. (2019). First assessment of phytoplankton diversity in a Marrocan shallow reservoir (Sidi Abderrahmane). *Saudi Journal of Biological Sciences, 26*(3), 431-438. doi.org/10.1016/j.sjbs.2017.11.047

Berger, W.H., and Parker, F.L. (1970). Diversity of planktonic foraminifera in deep-sea sediments. *Science, 168,* 1345– 1347.

Bose, R., Bose, A.K., Das, A.K., Parashar, A., and Roy, K. (2019). Fish diversity and limnological parameters influencing fish assemblage pattern in Chambal River Basin of Madhya Pradesh, India. *Proceedings of the National Academy of Sciences, India Section B: Biological Sciences, 89*(2), 461-473. https: //doi.org/10.1007/s40011-017-0958-5

Brönmark, C., and Hansson, L.A. (2002). Environmental issues in lakes and ponds: Current state and perspectives. *Environmental Conservation, 29*(3), 290-307. https: //doi.org/10.1017/S0376892902000218

Catherine, Q., Susanna, W., Isidora, E.S., Mark, H., Aurelie, V., and Jean-François, H. (2013). A review of current knowledge on toxic benthic freshwater cyanobacteria–ecology, toxin production and risk management. *Water Research, 47*(15), 5464-5479. http: //dx.doi.org/10.1016/j.watres.2013.06.042

Çelekli, A., and Sahin, G. (2021). Bio-assessment of wastewater effluent conditions with algal pollution index and multivariate approach. *Journal of Cleaner Production, 310*, 1-11.

Chen, J., and Lu, J. (2014). Effects of land use, topography and socio-economic factors on river water quality in a mountainous watershed with intensive agricultural production in East China. *PloS one, 9*(8), 1-12. https: //doi.org/10.1371/journal.pone.0102714

Chen, R., Ju, M., Chu, C., Jing, W., and Wang, Y. (2018). Identification and quantification of physico-chemical parameters influencing chlorophyll-a concentrations through combined principal component analysis and factor analysis: A case study of the Yuqiao reservoir in China. *Sustainability, 10*(4), 1-15. doi: 10.3390/su10040936

Cheraghpour, J., Afsharzadeh, S., Sharifi, M., Ghadi, R.R., and Masoudi, M. (2013). Phytoplankton diversity assessment of Gandoman Wetland, West of Iran. *Iranian Journal of Botany, 19*(2), 153-16.

Das, S., Nath, K., and Chowdhury, R. (2021). Comparative studies on biomass productivity and lipid content of a novel blue-green algae during autotrophic and heterotrophic growth. *Environmental Science and Pollution Research, 28*(10), 12107-12118. https: //doi.org/10.1007/s11356-020-09577-4

Desikachary, T.V. (1959). Cyanophyta. Indian Council for Agricultural Research, New Delhi.

Devercelli, M., and O'Farrell, I. (2013). Factors affecting the structure and maintenance of phytoplankton functional groups in a nutrient rich lowland river. *Limnologica, 43*(2), 67-78. http: //dx.doi.org/10.1016/j.limno.2012.05.001

Du, X., Liu, H., Yuan, L., Wang, Y., Ma, Y., Wang, R., Chen, X., Losiewicz, M.D., Guo, H., and Zhang, H. (2019). The diversity of cyanobacterial toxins on structural characterization, distribution and identification: A systematic review. *Toxins, 11*(9), 1-34. doi: 10.3390/toxins11090530

El-Kassas, H.Y., and Gharib, S.M. (2016). Phytoplankton abundance and structure as indicator of water quality in the drainage system of the Burullus Lagoon, southern Mediterranean coast, Egypt. *Environmental Monitoring and Assessment, 188*(9), 1-14. DOI 10.1007/s10661-016-5525-7

Elliott, J.A. (2012). Is the future blue-green? A review of the current model predictions of how climate change could affect pelagic freshwater cyanobacteria. *Water Research*, *46*(5), 1364-1371.

Fang, S., Del Giudice, D., Scavia, D., Binding, C.E., Bridgeman, T.B., Chaffin, J.D., Evans, M.A., Guinness, J., Johengen, T.H., and Obenour, D.R. (2019). A space-time geostatistical model for probabilistic estimation of harmful algal bloom biomass and areal extent. *Science of the Total Environment*, *695*, 1-12.

Fetahi, T., Schagerl, M., and Mengistou, S. (2014). Key drivers for phytoplankton composition and biomass in an Ethiopian highland lake. *Limnologica*, *46*, 77-83. http: //dx.doi.org/10.1016/j.limno.2013.10.007

Ghobara, M., and Salem, Z.E. (2017). Spatiotemporal fluctuations in phytoplankton communities and their potential indications for the pollution status of the irrigation and drainage water in the middle Nile Delta area, Egypt. *In Conventional Water Resources and Agriculture in Egypt*, *74*, 317-345. DOI 10.1007/698_2017_185

Gogoi, P., Sinha, A., Sarkar, S.D., Chanu, T.N., Yadav, A.K., Koushlesh, S.K., Borah, S., Das, S.K., and Das, B.K. (2019). Seasonal influence of physico-chemical parameters on phytoplankton diversity and assemblage pattern in Kailash Khal, a tropical wetland, Sundarbans, India. *Applied Water Science*, *9*(7), 1-13.

He, G., Fang, H., Bai, S., Liu, X., Chen, M., and Bai, J. (2011). Application of a three-dimensional eutrophication model for the Beijing Guanting Reservoir, China. *Ecological Modelling*, *222*(8), 1491-1501.

Hussien, N.R., and Zaghloul, F.A. (2017). Assessment of Phytoplankton Community as a bio indicator of organic pollution in the drainage system of the Edku Lagoon, Egypt. *Asian Journal of Advanced Basic Sciences*, *5*(1), 59-71.

Ji, D., Wells, S.A., Yang, Z., Liu, D., Huang, Y., Ma, J., and Berger, C.J. (2017). Impacts of water level rise on algal bloom prevention in the tributary of three Gorges Reservoir, China. *Ecological Engineering*, *98*, 70-81.

Jiang, Y.J., He, W., Liu, W.X., Qin, N., Ouyang, H.L., Wang, Q.M., Kong, X.Z., He, Q.S., Yang, C., Yang, B., and Xu, F.L. (2014). The seasonal and spatial variations of phytoplankton community and their correlation with environmental factors in a large eutrophic Chinese lake (Lake Chaohu). *Ecological Indicators*, *40*, 58-67. http: //dx.doi.org/10.1016/j.ecolind.2014.01.006

Jose, L., and Kumar, C. (2011). Evaluation of pollution by Palmer's algal pollution index and physico-chemical analysis of water in four temple ponds of Mattancherry, Ernakulam, Kerala. *Nature, Environment and Pollution Technology*, *10*(3), 471-472.

Kantachote, D., Dangtago, K., and Siriwong, C. (2009). Treatment efficiency in wastewater treatment plant of Hat Yai Municipality by quantitative removal of microbial indicators. *Songklanakarin Journal of Science and Technology*, *31*(5), 567-576.

Ke, Z., Xie, P., and Guo, L. (2008). Controlling factors of spring–summer phytoplankton succession in Lake Taihu (Meiliang Bay, China). *Hydrobiologia, 607*(1), 41-49.

Kim, S., Chung, S., Park, H., Cho, Y., and Lee, H. (2019). Analysis of environmental factors associated with cyanobacterial dominance after river weir installation. *Water, 11*(6), 1-24. doi: 10.3390/w11061163

Kiran, M. T., Bhaskar, M. V., and Tiwari, A. (2016). Phycoremediation of eutrophic lakes using diatom algae. *Lake Sciences and Climate Change*, 103-115. DOI: 10.5772/64111

Kireta, A.R., Reavie, E.D., Sgro, G.V., Angradi, T.R., Bolgrien, D.W., Hill, B.H., and Jicha, T.M. (2012). Planktonic and periphytic diatoms as indicators of stress on great rivers of the United States: Testing water quality and disturbance models. *Ecological Indicators, 13*(1), 222-231.

Kouhanestani, Z.M., Roelke, D.L., Ghorbani, R., and Fujiwara, M. (2019). Assessment of spatiotemporal phytoplankton composition in relation to environmental conditions of Gorgan Bay, Iran. *Estuaries and Coasts, 42*(1), 173-189.

Kozak, A., Budzyñska, A., Dondajewska-Pielka, R., Kowalczewska-Madura, K., and Go^{3}dyn, R. (2020). Functional groups of phytoplankton and their relationship with environmental factors in the restored Uzarzewskie Lake. *Water, 12*(2), 1-14. doi: 10.3390/w12020313

Krupa, *E.G.*, Barinova, S.S., Romanova, S.M., and Malybekov, A.B. (2016). Hydrobiological assessment of the high mountain Kolsay lakes (Kungey Alatau, Southeastern Kazakhstan) ecosystems in climatic gradient. *British Journal of Environment and Climate Change*, 6(4), 259-278.

Kshirsagar, A.D. (2013). Use of algae as a bioindicator to determine water quality of river Mula from Pune city, Maharashtra (India). *Universal Journal of Environmental Research and Technology, 3*(1), 79-85.

Kumar, P., and Mahajan, A.K. (2020). Trophic status and its regulating factor determination at the Rewalsar Lake, northwest Himalaya (HP), India, based on selected parameters and multivariate statistical analysis. *SN Applied Sciences, 2*(7), 1-11. https: //doi.org/10.1007/s42452-020-3082-8

Lavoie, I., Campeau, S., Zugic-Drakulic, N., Winter, J.G., and Fortin, C. (2014). Using diatoms to monitor stream biological integrity in Eastern Canada: An overview of 10 years of index development and ongoing challenges. *Science of the Total Environment, 475*, 187-200.

Le, T.T.H., Zeunert, S., Lorenz, M., and Meon, G. (2017). Multivariate statistical assessment of a polluted river under nitrification inhibition in the tropics. *Environmental Science and Pollution Research, 24*(15), 13845-13862. DOI 10.1007/s11356-017-8989-2

Li, H., Fan, Y., Gong, Z., and Zhou, D. (2020). Water accessibility assessment of freshwater wetlands in the Yellow River Delta National Nature Reserve, China.

Ecohydrology and Hydrobiology, 20(1), 21-30. https: //doi.org/10.1016/j.ecohyd.2019.02.006

Lu, X., Tian, C., Pei, H., Hu, W., and Xie, J. (2013). Environmental factors influencing cyanobacteria community structure in Dongping Lake, China. *Journal of Environmental Sciences, 25*(11), 2196-2206. DOI: 10.1016/S1001-0742(12)60297-6

Lv, J., Wu, H., and Chen, M. (2011). Effects of nitrogen and phosphorus on phytoplankton composition and biomass in 15 subtropical, urban shallow lakes in Wuhan, China. *Limnologica, 41*(1), 48-56. doi: 10.1016/j.limno.2010.03.003

Mahapatra, D.M., Chanakya, H.N., and Ramachandra, T.V. (2011). Assessment of treatment capabilities of Varthur Lake, Bangalore, India. *International Journal of Environmental Technology and Management, 14*(1/2/3/4), 84-102.

Mahapatra, D.M., Chanakya, H.N., and Ramachandra, T.V. (2013). *Euglena* sp. as a suitable source of lipids for potential use as biofuel andsustainable wastewater treatment. *Journal of Applied Phycology, 25*(3), 855–865. https: //doi.org/10.1007/s10811-013-9979-5.

Marchello, A.E., Lombardi, A.T., Dellamano-Oliveira, M.J., and de Souza, C.W. (2015). Microalgae population dynamics in photobioreactors with secondary sewage effluent as culture medium. *Brazilian Journal of Microbiology, 46*(1), 75-84.

Merel, S., Walker, D., Chicana, R., Snyder, S., Baurès, E., and Thomas, O. (2013). State of knowledge and concerns on cyanobacterial blooms and cyanotoxins. *Environment International, 59*, 303-327. http: //dx.doi.org/10.1016/j.envint.2013.06.013

Miao, X., Wang, S., Liu, M., Ma, J., Hu, J., Li, T., and Chen, L. (2019). Changes in the phytoplankton community structure of the Backshore Wetland of Expo Garden, Shanghai from 2009 to 2010. *Aquaculture and Fisheries, 4*(5), 198-204. https: //doi.org/10.1016/j.aaf.2019.02.004

Mishra, P., Garg, V., and Dutt, K. (2019). Seasonal dynamics of phytoplankton population and water quality in Bidoli reservoir. *Environmental Monitoring and Assessment, 191*(3), 1-12.

Muthukumar, C., Muralitharan, G., Vijayakumar, R., Panneersevam, A., and Thajuddin, N. (2007). Cyanobacterial biodiversity from different freshwater ponds of Thanjavur, Tamil Nadu (India). *Acta Botanica Malacitana, 32*, 17-25.

Nandigam, J., Rangaiah, S.G., and Geddada, M.N.R. (2016). A study on seasonal changes of microalgae in relation to the physico-chemical parameters of Satyavaram Pond, Srikakulam Dist, India. *Indian Journal of Geo Marine Sciences, 45*(12), 1660-1668.

Narchonai, G., Arutselvan, C., LewisOscar, F., and Thajuddin, N. (2019). Deciphering the microalgal diversity and water quality assessment of two urban temple ponds in Pondicherry, India. *Biocatalysis and Agricultural Biotechnology, 22*, 1-9.

Noel, S.D., and Rajan, M.R. (2015). Evaluation of organic pollution by Palmer's algal genus index and physico-chemical analysis of Vaigai River at Madurai, India. *Natural Resources and Conservation, 3*(1), 7-10. DOI: 10.13189/nrc.2015.030102

Ntengwe, F.W. (2006). Pollutant loads and water quality in streams of heavily populated and industrialised towns. *Physics and Chemistry of the Earth, 31*(15-16), 832-839. https: //doi.org/10.1016/j.pce.2006.08.025

Paerl, H.W., Hall, N.S., and Calandrino, E.S. (2011). Controlling harmful cyanobacterial blooms in a world experiencing anthropogenic and climatic-induced change. *Science of the Total Environment, 409*(10), 1739-1745. doi: 10.1016/j.scitotenv.2011.02.001

Parvez, M.A., Uddin, M.M., Islam, M.K., and Kibria, M.M. (2019). Physico-chemical and biological monitoring of water quality of Halda river, Bangladesh. *International Journal of Environmental and Science Education, 14*(4), 169-181.

Prescott, G.W. (1970). How to Know Freshwater Algae.2nd ed., W.C. Brown. Dubuque, Iowa.

Qu, Z., Duan, P., Cao, X., Liu, M., Lin, L., and Li, M. (2019). Comparison of monoculture and mixed culture (*Scenedesmus obliquus* and wild algae) for C, N, and P removal and lipid production. *Environmental Science and Pollution Research, 26*(20), 20961-20968. https: //doi.org/10.1007/s11356-019-05339-z

Ramachandra T.V., Sincy V., and Asulabha K.S. (2020a). Efficacy of rejuvenation of lakes in Bengaluru, India. *Green Chemistry and Technology Letters, 6*(1), 14-26. doi.org/10.18510/gctl.2020.613

Ramachandra, T. V., Vinay, S., Asulabha, K. S., Sincy V., Bhat, S. P., Mahapatra, D.M., and Aithal, B. H. (2017). Rejuvenation blueprint for lakes in Vrishabhavathi Valley. *ENVIS Technical Report 122*, Energy and Wetlands Research Group, CES, Indian Institute of Science, Bangalore, 2017.

Ramachandra, T.V., and Aithal, B.H. (2015). Wetlands: Kidneys of Bangalore's landscape. *National wetlands, 37*, 12-16.

Ramachandra, T.V., Asulabha, K.S., and Lone, A.A. (2014). Nutrient enrichment and proliferation of invasive macrophytes in urban lakes. *Journal of Biodiversity, 5*(1-2), 33-44.

Ramachandra, T.V., Asulabha, K.S., Sincy, V., Vinay, S., Bhat, S.P., and Aithal, B.H. (2015). Sankey Lake: Waiting for an immediate sensible action. *ENVIS Technical Report 74*, CES, Indian Institute of Science, Bangalore 560012.

Ramachandra, T.V., Meera, D.S., and Alakananda, B. (2013). Influence of catchment land cover dynamics on the physical, chemical and biological integrity of wetlands. *Environment and We-International Journal of Science and Technology, 8*(1), 37-54.

Ramachandra, T.V., Sincy, V., Asulabha, K.S., Mahapatra, D.M., Bhat, S.P. and Aithal, B.H. (2018). Optimal treatment of domestic wastewater through constructed wetlands. *Journal of Biodiversity, 9*(1-2), 81-102. DOI: 10.31901/24566543.2018/09.1-2.077

Ramachandra, T.V., Sudarshan, P., Vinay, S., Asulabha, K.S., and Varghese, S. (2020b). Nutrient and heavy metal composition in select biotic and abiotic components of Varthur wetlands, Bangalore, India. *SN Applied Sciences, 2*(8), 1-14.

Ray, J.G., Santhakumaran, P., and Kookal, S. (2021). Phytoplankton communities of eutrophic freshwater bodies (Kerala, India) in relation to the physico-chemical water quality parameters. *Environment, Development and Sustainability*, 1-32. https: //doi.org/10.1007/s10668-019-00579-y

Salem, Z., Ghobara, M., and El Nahrawy, A.A. (2017).Spatio-temporal evaluation of the surface water quality in the middle Nile Delta using Palmer's algal pollution index. *Egyptian Journal of Basic and Applied Sciences, 4*(3), 219-226.

Saxena, A., Tiwari, A., Kaushik, R., Iqbal, H. M., and Parra-Saldívar, R. (2020). Diatoms recovery from wastewater: Overview from an ecological and economic perspective. *Journal of Water Process Engineering*, 39, 1-14. doi: 10.1016/j. jwpe.2020.101705

Shahare, P.C.A. (2017). Systematic account of the freshwater diatom from Chulband river of Gondia district, Maharashtra (India). *IJRBAT, 5*(1), 72-75.

Shams, M., and Shamsabadi, K.S. (2019). Identification of algae as pollution bioindicators in Shakh-Kenar, Gavkhouni Wetland, Isfahan. *Journal of Phycological Research, 3*(2), 386-394.

Shanthala, M., Hosmani, S.P., and Hosetti, B.B. (2009).Diversity of phytoplanktons in a waste stabilization pond at Shimoga Town, Karnataka State, India. *Environmental Monitoring and Assessment, 151*(1), 437-443.

Sharma, R.C., Singh, N., and Chauhan, A. (2016). The influence of physico-chemical parameters on phytoplankton distribution in a head water stream of Garhwal Himalayas: A case study. *The Egyptian Journal of Aquatic Research, 42*(1), 11-21. http: //dx.doi.org/10.1016/j.ejar.2015.11.004

Shil, S., Singh, U.K., and Mehta, P. (2019). Water quality assessment of a tropical river using water quality index (WQI), multivariate statistical techniques and GIS. *Applied Water Science, 9*(7), 1-21. https: //doi.org/10.1007/s13201-019-1045-2

Singh, S.P., and Singh, P. (2015). Effect of temperature and light on the growth of algae species: A review. *Renewable andSustainable Energy Reviews, 50*, 431-444. http: //dx.doi.org/10.1016/j.rser.2015.05.024

Su, M., Yu, J., Pan, S., An, W., and Yang, M. (2014). Spatial and temporal variations of two cyanobacteria in the mesotrophic Miyun reservoir, China. *Journal of Environmental Sciences, 26*(2), 289-298. DOI: 10.1016/S1001-0742(13)60433-7

Taipale, S.J., Vuorio, K., Aalto, S.L., Peltomaa, E., and Tiirola, M. (2019). Eutrophication reduces the nutritional value of phytoplankton in boreal lakes. *Environmental Research, 179*, 1-13. https: //doi.org/10.1016/j.envres.2019.108836

Teta, R., Romano, V., Della Sala, G., Picchio, S., De Sterlich, C., Mangoni, A., Tullio, G.D., Costantino, V., and Lega, M. (2017). Cyanobacteria as indicators of water quality in Campania coasts, Italy: a monitoring strategy combining remote/ proximal sensing and *in situ* data. *Environmental Research Letters, 12*(2), 1-12.

Tian, C., Pei, H., Hu, W., and Xie, J. (2012). Variation of cyanobacteria with different environmental conditions in Nansi Lake, China. *Journal of Environmental Sciences, 24*(8), 1394-1402. DOI: 10.1016/S1001-0742(11)60964-9

Udhayakumar, R., Manivannan, P., Raghu, K., and Vaideki, S. (2016). Assessment of physico-chemical characteristics of water in Tamil Nadu. *Ecotoxicology and Environmental Safety, 134*, 474-477. http: //dx.doi.org/10.1016/j.ecoenv.2016.07.014

Varol, M., 2020. Use of water quality index and multivariate statistical methods for the evaluation of water quality of a stream affected by multiple stressors: A case study. *Environmental Pollution, 266*, 1-10.

Vishal, R., and Meeta, B. (2020). Algal indices as a biomonitoring tool to assess eutrophication in the urban ponds: A case study. *Pollution, 6*(4), 751-757.

Wang, X., Wang, Y., Liu, L., Shu, J., Zhu, Y., and Zhou, J. (2013). Phytoplankton and eutrophication degree assessment of Baiyangdian Lake Wetland, China. *The Scientific World Journal, 2013*, 1-9. https: //doi.org/10.1155/2013/436965

Wijeyaratne, W.M.D.N., and Nanayakkara, D.B.M. (2020). Monitoring of water quality variation trends in a tropical urban wetland system located within a Ramsar wetland city: A GIS and phytoplankton based assessment. *Environmental Nanotechnology, Monitoring and Management, 14*, 1-10.

Wu, Y., Li, L., Gan, N., Zheng, L., Ma, H., Shan, K., Liu, J., Xiao, B., and Song, L. (2014). Seasonal dynamics of water bloom-forming Microcystis morphospecies and the associated extracellular microcystin concentrations in large, shallow, eutrophic Dianchi Lake. *Journal of Environmental Sciences, 26*(9), 1921-1929. http: //dx.doi.org/10.1016/j.jes.2014.06.031

Xu, C., Poon, K., Choi, M.M., and Wang, R. (2015). Using live algae at the anode of a microbial fuel cell to generate electricity. *Environmental Science and Pollution Research, 22*(20), 15621-15635. DOI 10.1007/s11356-015-4744-8

Xu, G., Li, P., Lu, K., Tantai, Z., Zhang, J., Ren, Z., Wang, X., Yu, K., Shi, P., and Cheng, Y. (2019). Seasonal changes in water quality and its main influencing factors in the Dan river basin. *Catena, 173*, 131-140. https: //doi.org/10.1016/j.catena.2018.10.014

Yang, J., Lv, H., Liu, L., Yu, X., and Chen, H. (2016). Decline in water level boosts cyanobacteria dominance in subtropical reservoirs. *Science of the Total Environment, 557*, 445-452.

Yang, J., Yu, X., Liu, L., Zhang, W., and Guo, P. (2012). Algae community and trophic state of subtropical reservoirs in southeast Fujian, China. *Environmental Science and Pollution Research, 19*(5), 1432-1442. DOI 10.1007/s11356-011-0683-1

Yilmaz, N., Yardimci, C. H., Elhag, M., and Dumitrache, C. A. (2018). Phytoplankton Composition and Water Quality of Kamil Abdu° Lagoon (Tuzla Lake), Istanbul-Turkey. *Water, 10*(5), 1-13. doi: 10.3390/w10050603

Zhang, M., Yu, Y., Yang, Z., and Kong, F. (2016). Deterministic diversity changes in freshwater phytoplankton in the Yunnan–Guizhou Plateau lakes in China. *Ecological Indicators, 63,* 273-281. http: //dx.doi.org/10.1016/j.ecolind.2015.12.017

Zhao, G., Gao, X., Zhang, C., and Sang, G. (2020). The effects of turbulence on phytoplankton and implications for energy transfer with an integrated water quality-ecosystem model in a shallow lake. *Journal of Environmental Management, 256,* 1-11. https: //doi.org/10.1016/j.jenvman.2019.109954.

Chapter 11

Biodiversity Management of most Impacted Soil Earthworms and their Karyotpyes

Jayashree M. Wodeyar[1] *and Shankerappa Hatti*[2]

[1]*Department of Zoology, Sheshadripuram College, Yelahanka New Town, Bengaluru, Karnataka*
[2]*Department of UG and PG Studies and Research in Zoology, Government College (Autonomous), Kalaburagi – 585 106, Karnataka*
e-mail: geethamalimath@gmail.com

Biodiversity of earthworms are the oldest, significant, Environmental indicators of soil health, organic practices studied among all other studies (Haughley farm 1980/81) which could be compared between earthworm populations and soil properties. Double the population and diversity obtained when compared with farming method. Loss, Climate change, limited hygiene, and a variety of daily activities produces about 38 billion metric tons of organic, solid and hazardous wastes/year. As this unusable waste if converted into useful and sustainable nutrient rich agricultural product with the key drivers of mineral degradation mediated by earthworms plays an important role in ecosystem processes not only through their positive effects on soil structure, but also through their promotion in preventing and monitoring environmental pollution under a broad spectrum of Vermicomposting Technology. Vermicompost is the mineralized and humified organic matter produced by the Earthworms. The nature's gift for utilization of Vermicompost which results in recycling of solid/organic/ hazardous waste. Monoculture and polyculture practice and compared to obtain maximum growth and development. The gut of Earthworm is known to accumulate metals. Surprisingly earthworms have been an important experimental model since the very beginning of immunology. The oligochaete annelids; are harmless ecosystem's engineers, contributing to a wide range of nutrient cycling and geochemical processes in the ecosystem and are found globally having greatest impact on health and environment.

Karyological studies and Chromosome analysis is an essential part of systematic studies of any group of organisms, and can play an important role, especially in taxonomically complicated genera so also the karyology of Earthworms is least studied. Hence highlighting the karyological studies of earthworms in Indian agricultural soil not only in changing environment but also its impact on health/organic/solid/hazardous waste management. In the last four decades important cellular and humoral pathways were described and numerous biologically active compounds were characterized and often cloned. Screening of earthworm immunological parameters as markers/indicators of impaired environmental conditions.

The methods used are gonads (Meiotic stages), tail tissue (Mitotic stages) of Eisenia fetida (2N=22, N=11) and Pheritima elongata (2N=100, N=50) treating with hypotonic solution (0.5 per cent sodium citrate sol.) Fixed in fixative. Air dried technique, stages resulted in getting the chromosome number of the said above species.

Keywords: *Biodiversity, Management, Vermiculture, Earthworms and their Karyotype.*

Introduction

Biodiversity of earthworms are varied live in decaying matter, conifer logs, mud, arboreal, aquatic, saline water and soil derived from serpentine. Earthworms biodiversity is the oldest, significant, Environmental indicators of soil health, organic practises studied among all other studies (Haughley farm 1980/81) which could be compared between earthworm populations and soil properties. Double the population and diversity obtained when compared with farming method. The impact of earthworms on Environment are Biological, chemical, and physical. Loss, Climate change, limited hygiene, and a variety of daily activities produces about 38 billion metric tons of organic, solid and hazardous wastes/year. As this unusable waste if converted into useful and sustainable nutrient rich agricultural product with the key drivers of mineral degradation mediated by earthworms plays an important role in ecosystem processes not only through their positive effects on soil structure, but also through their promotion in preventing and monitoring environmental pollution under a broad spectrum of Vermicomposting Technology. Vermicompost is the mineralized and humified organic matter produced by the Earthworms. The nature's gift for utilization of Vermicompost which results in recycling of solid/organic/hazardous aste. The gut of Earthworm is known to accumulate metals. Surprisingly earthworms have been an important experimental model since the very beginning of immunology. The oligochaete annelids; are harmless ecosystem's engineers, contributing to a wide range of nutrient cycling and geochemical processes in the ecosystem and are found globally having greatest impact on health and environment. Decomposition efficiency of different species of earthworms in vermireactors. Earthworms are three types; they are Anecic, Epigeic and Endogeic. An antibacterial activity of coelomic fluid of earthworms is against highly pathogenic bacteria. As these pathogenic bacteria are known to produce at least one common antigen strain which is also present in sheep RBCs. It is found that in the soil selected in one of the village Chowdayadanapura, the agriculture land where vermicompost is used some of the earthworm species especially *Eisenia foetida* apart from encouraging the establishment of beneficial microorganisms can also inhibit soil borne pathogens.

Much of the emphasis is being given to organic farming in current years. The current study was to test which speices is more beneficial for the Indian Agricultural soil using three different species they are 1. *Eisenia foetida*. 2. *Polyphertima elongata* and *Eudrilus euginiae*. Among these three, Polypheretima a peregrine in Indian regions are less used than Eudrilus euginiae commonly utilized in vermiculture and vermitechnology (J M Julka).

The third one that is Eugenia fetida, an European peregrine species is found to be more useful both in terms of establishment of beneficial microorganisms and most suitable for many different types of crops growth in providing high nutritional quality soil in this selected village. May be due to its least number of chromosomes.

Methods and Materials

Figure 11.1: Earthworms Cultured in a Vermireactor.

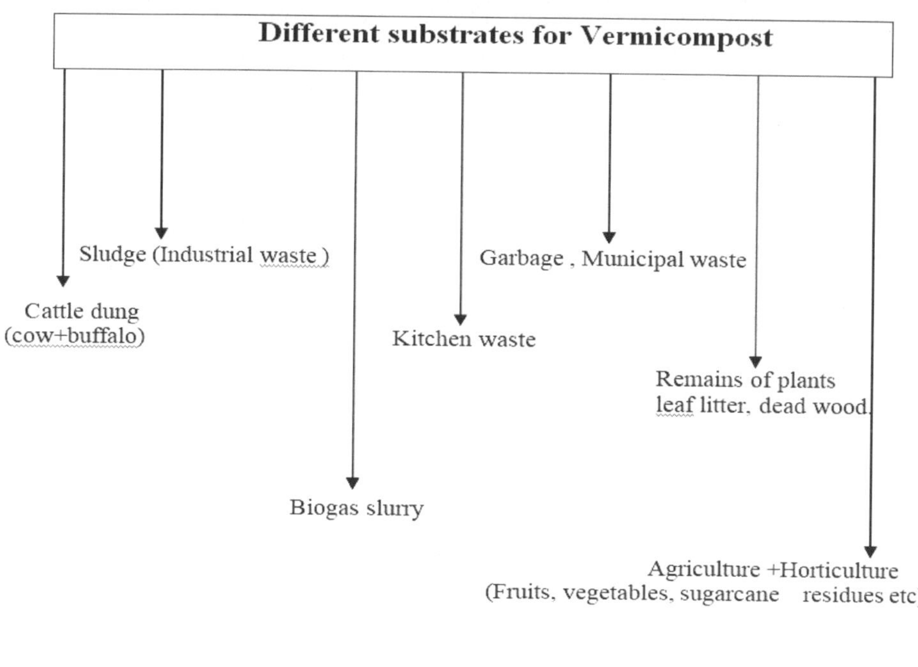

Parameters	Monoculture with epigeic	Monoculture with anecic	Polyculture with both epigeic and anecic
pH	8.5 ± 0.02a	8.4 ±0.01 a	8.5±0.03
Corg (g kg -1)	281± 5.4a	278 ±4.1 a	261 ±3.7 a
N tot (g kg -1)	7.56 ±0.03 a	6.73± 0.02 a	9.23 ±0.02 b
NH4-N (mg kg -1)	25. 6 ±0.7 a	28. 4 ±0.6 b	40. 9 ±0.8 c
NO3-N (mg kg -1)	363. 1 ±1.7 a	480 .4 ±4.8 b	721 ±4.7 c
P avail (g kg -1)	5. 98± 0.02 a	2.83± 0.02 a	2.55 ±0.01 a
Kexch (gkg-1)	5. 98 ±0.02 a	7.00 ±0.01 a	6.16 ±0.03 b
C:N ratio	37 .4 ±0.65 a	36.2 ±0.78 b	30. 11 ±0.28 b

Monoculture and Polyculture Comparative Practices to Obtain Maximum Yield Quantity/Quality-Wise

1. Selection of suitable species of earthworms for both monoculture and polyculture.
2. Decomposition efficiency of monoculture and polyculture of different species of earthworms.
3. Solid/hazardous waste management is determined
4. Significance of Vermiwash was also carried out

Different quantities of earthworms are taken and mixed (inoculated) in organic wastes and earthworms' response in terms growth and reproduction was monitored. The biomass produced formation of cocoon, weight of the worms, size and length of the earthworms. Earthworms' numbers and other factors such as temperature, feed, stock pH, humidity *etc.* have been monitored.

Karyological studies of different species were carried in the following method. The methods used are gonads (Meiotic stages), tail tissue (Mitotic stages) of *Eisenia fetida* (2N=22, N=11) and *Pheritima elongata* (2N=100, N=50) treating with hypotonic solution (0.5 per cent sodium citrate sol.) Fixed in fixative. Air dry technique was used for chromosomal analysis, different stages of mitotic division were observed under zoom microscope. Stages resulted in getting the chromosome number of the said above species

Karyotype and Idiogram of the *Eisenia foetida*

Idiogram Prepared on the Bases of Decreasing Length of Chromosome Pair Number from One (1) to Eleven (11)

Chromosome Pair Number	*Absolute Length of (p+q) of the Chromosome*
1	2.32
2	2
3	1.83

Chromosome Pair Number	*Absolute Length of (p+q) of the Chromosome*
4	1.76
5	1.63
6	1.56
7	1.45
8	0.94
9	0.64
10	0.51
11	0.33

Histogram of the Ideogram Prepared on the Bases of Decreasing Length of Chromosome Pair Numbers from One to Eleven

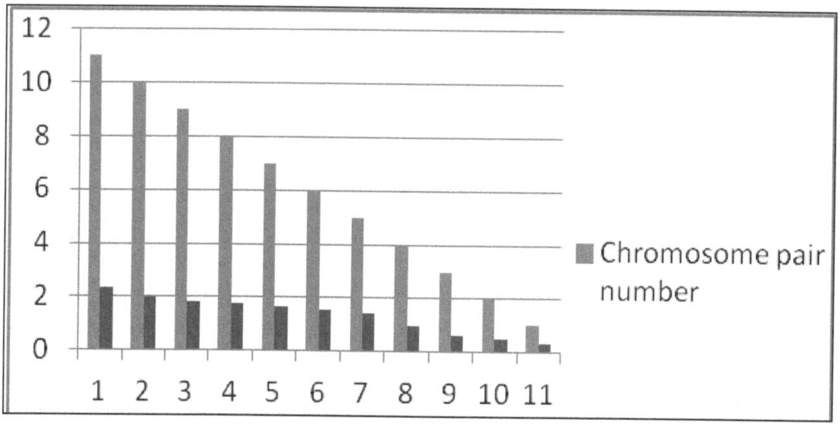

Histogram of Morphometric Analysis of *Eisenia foetida* Species.

Conclusion

Earthworms have become crucial drivers of the process in fragmenting and conditioning the substrate increasing surface area for micro biological activities of the micro organisms which are responsible for the bio chemical degradation of the organic matter. It was seen that "Permaculture" earthworms role in acting as an innumerable army of pistons pumping air in and out os soil on a 24 hours cycle (more during night) thus modified vital organic components that made soil healthy (Bill Mollison). Micro flora and micro fauna population were affected by earthworms directly or indirectly under the following mechanisms of a) Comminution. b) Burrowing and casting, c) Grazing and Dispersal. This changed substrates physiochemical and Biological states and caused drastic shift in the density, diversity, structure and activity of microbial communities within the Drilosphere.

Morphometric Analysis of the Karyotype of *Eisenia foetida* (N=11, 2N=22)

Chromosome Pair Number	*Mean Length of Short Arm (p) in um.*	*Mean Length of Long Arm (q) in um*	*Absolute Length (p+q) of the Chromosome*	*Arm Ratio q/p*	*Total Complement Length per cent*	*Centro-Meric Index*	*Nomenclature*
1	0.86	1.46	2.32	1.69	7.68	37.06	Submetacentric
2	0.64	1.36	2.00	2.12	6.62	32.00	Submetacentric
3	0.66	1.10	1.76	1.66	5.83	37.50	Submetacentric
4	0.61	1.22	1.83	2.00	6.06	33.33	Metacentric
5	0.48	1.08	1.56	2.25	5.16	30.76	Metacentric
6	0.22	0.72	0.94	3.27	3.11	23.40	Subtelocentric
7	0.41	1.04	1.45	2.53	4.80	28.27	Subtelocentric
8	0.39	1.24	1.63	3.17	5.41	23.92	Subtelocentric
9	0.08	0.54	0.64	6.75	2.12	12.50	Telocentric
10	0.05	0.46	0.51	9.20	1.68	09.80	Telocentric
11	0.02	0.31	0.33	0.15	1.09	06.06	Acrocentric

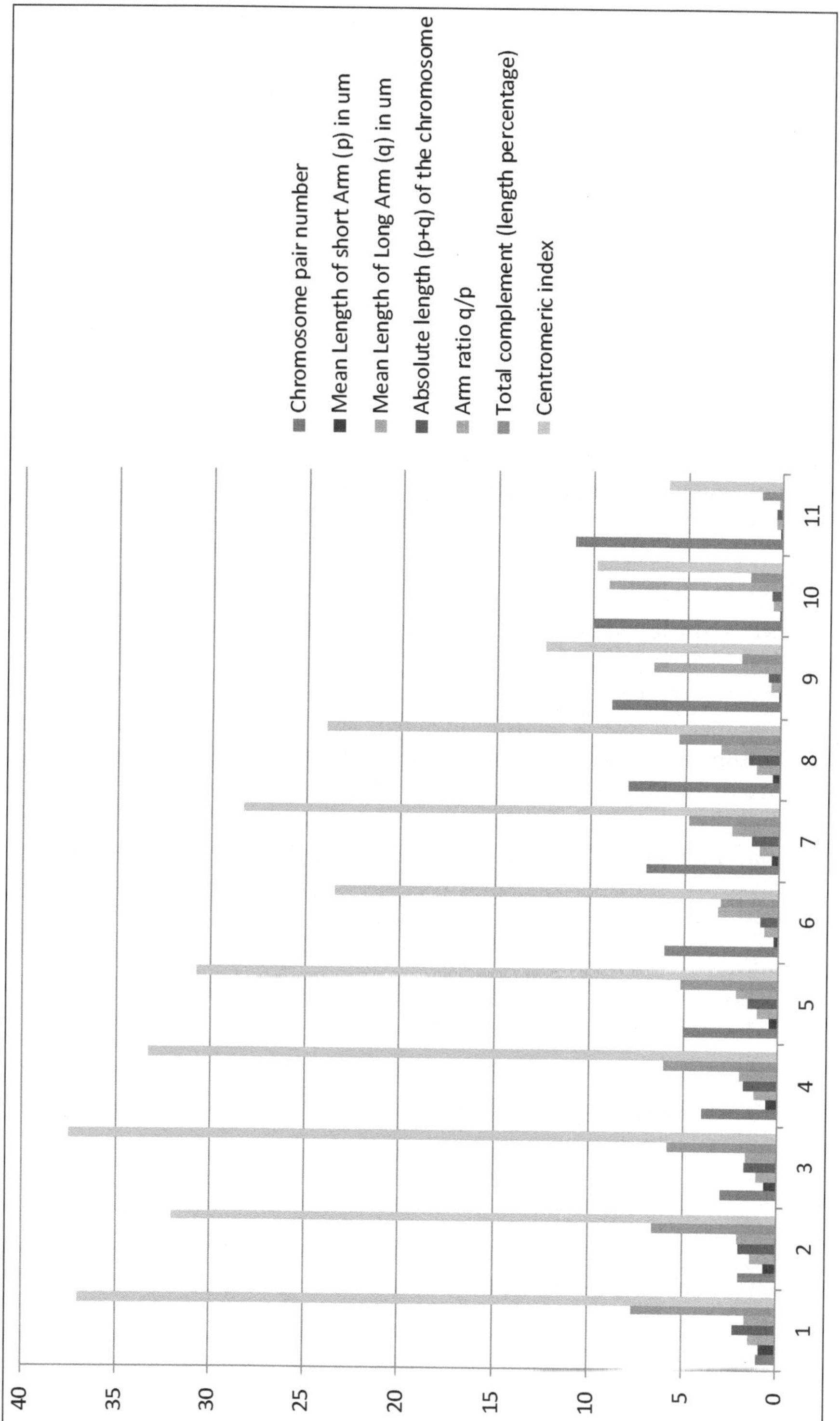
Chromosome pair number
Mean Length of short Arm (p) in um
Mean Length of Long Arm (q) in um
Absolute length (p+q) of the chromosome
Arm ratio q/p
Total complement (length percentage)
Centromeric index
0
5
10
15
20
25
30
35
40
1
2
3
4
5
6
7
8
9
10
11

The microbial load of vermireactors was measured through substrate- induced respiration rate (SIR), microbial biomass N content and rate of dehydrogenase activity, while mineralization rate was evaluated measuring some chemical parameters of the substrate. Polyculture is more beneficial to accelerate the mineralization rate in decomposing system. Monoculture is taken as experimental control.

It is concluded that significant change in chemical quality of substrate is being observed in polyculture. Vermicompost removes the problem of disposal of agriculture wastes, can be used as Biofertilizers and finally provides economical animal feed protein for fish and poultry Industries worldwide (Edwards 1985, Kale 2000). Thus solid/hazardous waste can be converted into more nutritive material both quantity/quality wise. Polyculture show N is enhanced 2-3 times more, P-1.3 - 1.6 times and K- 1.9- 2.8 tomes more. Polyculture includes 70 per cent of cowdung + organic sludge, after 77days there is reduction in C:N, increase in NPK, Calcium is seen in this vermreactors. This study also showed reduction in toxic metals such as cadmium, chromium, lead, nickel and zinc from the municipal waste collected from the dump yard from the near by government school and BO office from this village.

Thus It is concluded that polyculture vermireactor was more beneficial than the monoculture. secondly the chromosomal study conducted revealed that lesser the chromosomal number *Eisenia fetida* gave more cast, more biomass was obtained by another species that is *Eudrilus euginea*. Therefore it can be concluded that solid/hazardous waste can be converted into a very good compost containing rich valve in nutritional valve and degree of humification at the same time indicating different species karyology definitely effecting the nature of cast, production, growth, development and compost with high nutritive valve with humification.

Bulleted Summary

1. Different types of waste management through vermiculture.
2. Monoculture and polyculture practice to obtain maximum growth and development.
3. Production of rich nutritive valve compost.
4. Karyological studies of different species evaluate the biological, physiological, physiochemical parameters.
5. Morphometric analysis and histogram is studied for the quality and quantity production
6. Significance of vermiwash sprayed on different crops, yield was better than without vermiwash treatment.

REFERENCES

Catalogue of life: Eisenia fetida (Savigny, 1826).

Earthworm - The environmental Engineers: Review of vermiculture technologies for Environmental management. And resource development: Rajiv K Singh, Sunil Heart, Daldukh valani.

Earthworms andVermitechnolgy- Areview-Satyawati Sharma-Ashwini kumar, Anilpratap singh.

Edwards CA and Lofty JR Nitrogenous fertilizers and earthworm population in agricultural soul, soil Biology and Biochemistry, 1982, 14, 515-521.

Gandhi M, Sangwan V, Kapoor KK and Dilbaghi N, composting of household wastes with and without earthworms, *Environment* and *Ecology*, 1997, 15(2), 432-434.

Growth of the Earthworm Eisenia fetida in relation to population density and food rationing.Oikos 35 (1): 93-98.

Kale RD, Mallesh BC, Bano K and Bagyaraj DJ, InfluenceofVermicompost applications on the available macro nutrients and selected microbial population in a paddy field, *soil Biology and Biochemistry*, 1992, 24, 1317-1320.

Kitturmath MS, Giraddi RS and Shankarmurthy, Management of distillery waste using the eugine earthworm, *Eudrillus eigeniae* (kingberg) pest management *Econ. Zool*, 2005 13, 203-206.

Molecular phylogenetic study of earthworm *Eudrilus eugeniae*, International journal of Advanced Biotechnology and research 5, 202-207.

Nair J. Sekiozoic V and Anda M Effect of pre-composting on vermicomposting of kitchen waste, Bio resource Technology, 2006, 97, 2091-2095.

Role of earthworms in different organic waste–Asha, Ashok, Prafulla Soni, Ashotosh. Rao. R.K. Lakshmi-et al.

Tara crescent vermicomposting. Development Alternatives (DA) Sustainable livelihoods, (http: /www. Dainet.org/livelihood/default.htm) 2003.

Venkatachalaiah BK, The role of Karnataka compost development corporation Ltd. In: Organic farming Abstract, National seminar on organic farming and sustainable agriculture, October 9-11 APOF, Bangalore 1996, 24-25.

Vermicomposting in solid waste mangt.: Areview, Manyuchi M M and Phiri.

Vermicomposting of coffee pulp using the earthworm Eisenia fetida: Effects on C and N contents and the availability of nutrients" Biology and fertility of soils 22 (1-2): 162-166.

Chapter 12

Impact of Human Activity on Environment

Arunkumar Lagashetty

Department of Studies in Chemistry,
Vijayanagara Sri Krishnadevaraya University, Ballari, Karnataka
e-mail: arun.lagashetty@gmail.com

The contents of the environment are mainly matters the health issues of the living beings on environment. Scoping reviews have already been used to examine a variety of health related issues. Pollution is a major concern of new civilized world, which has a serious toxicological impact on human health and the environment. Most of these modern pollutants are rarely present in excessively large concentrations, so effects on health are usually far from immediate or obvious. The progressively larger number of people exposed to environmental nevertheless means that even small increases in relative risk can add up to major public health concerns. Long and short term exposure to harmful pollutants has a different toxicological impact on human health and causes many deceases. Air pollution, water pollution and soil pollution are considered as the major environmental risk factor in the incidence and progression of some chronic diseases and also impact badly on the environment. In present report, it is a part of discussion especially on toxicology of some air pollution like agriculture activity, mining activity etc. and their impact on human health. In addition to this some of the polluting sources and its health issues are also discussed. It is also tried to put some of the possible remedies reduce the environmental pollution are also envisaged.

Keywords: *Environment, Impact, Toxicology, Pollution.*

Introduction

Human activities have a negative impact on biological diversity due to globalization especially by industrialization and transportation. The use of pesticides and herbicides, and the release of some toxic compounds by chemical activity into the environment cause decrease biodiversity due its poisonous nature (Ali and Khan, 2017). Activities of every human being are governed by the resources available to them and interaction among species is quite common. The waste product of human may form the food supply of another species. Humans have the potential to gather resources from beyond their immediate surroundings and process them into different and more versatile forms. As a result of which the man made pollutants have overloaded the system, and the natural equilibrium is disturbed. Developmental activities have been associated with exploitation of many natural resources has resulted in fast depletion of energy resources like coal and petroleum, and minerals. Construction, mining activities and industrialization have interfered with the ecological balance of nature due to large scale impact. Natural resources like air, water and light are basic needs for every human for their survival. These processed resources were readily available in the biosphere, and the residues produced by the use of these resources were generally compatible with easily assimilated by the environment (Ahmad and Mujahid, 2011; Jeffrey *et al.*, 1997). Humans were better able than ever to satisfy their needs for air, water, food and shelter. In addition to this their attention is towards for comfortable survival. These acquired needs are usually met by items that must be processed or prepared. The production, distribution and use of such items usually results in more complex residuals and/or wastes, many of which are not compatible with or readily assimilated by the environment. The needs increase so will the complexity of the production chain, and mass and complexity of the pollutants generated (Mark *et al.*, 2019; Kiran *et al.*, 2013). To our knowledge, few studies have systematically reviewed the literature in the broad field of pollution exposure system, especially with regard to related health outcomes. This gap is filled by conducting a comprehensive scoping review of the literature with a focus on health outcomes affected by anthropogenic pollution. The purposes of this study for systematic overview of pollution sources and identify the different types of air pollution sources with its impact of human health.

Effect of Anthropogenic Pollution on Environment

Man made pollution is the main source for environmental pollution and it impacts lot on the environment. Some of the essential requirements for the comfortable leaving of the humans are major sources for environmental pollution. Agricultural activity, mining activity, transportation activity, standard leaving systems, housing activity *etc.* changes constituents of the environment (Díaz *et al.*, 2020). Some of the activities which badly effect the environment are discussed in the following section

Effects of Agriculture Activities

Agriculture is the food source for every living being on the world and it is one of the source environmental pollution. It effect on the area where the much agriculture takes place by the agricultural site, typically downstream and downwind. Broadly agricultural effects on environment are classification as local change, regional change and global change. Local change takes place near the forming site. One of the local change examples includes fertilizers sedimentation cause eutrophication. Regional change happens due to the combined effects of farming practices in the same large region. Deforestation, desertification, large scale pollution, increases in sedimentation in major rivers are the examples of regional change effects. Global change includes climatic changes as well as initially extensive changes in chemical cycles. The slash and burn of trees in forests to clear the land for cultivation and frequent shifting result in loss of forest cover. Clearing of forest cover exposes the soil to wind, rain and storms, thereby resulting in loss of top fertile layer of soil. During burn the organic matter in the soil gets destroyed and most of the nutrients are taken up by the crops within a short period, thus making the soil nutrient poor which forces the cultivators shift to another area (Kansal and Singh, 1983; Kloke, 1982). Recent agriculture method includes the use of many tools and chemicals; hence these practices have both many effects on environment. Pesticides have created a revolution in agriculture but the long term use of these chemicals has proved many effects. The major problems that have aroused due to the modern agricultural practices are related to fertilizers, pesticides, water-logging and salination (Chen *et al.*, 2015; Hey *et al.*, 2016).

Effects of Fertilizers

Almost fertilizers are chemical in natures which are used in modern agriculture system. Simply fertilizers have nitrogen, phosphorus and potassium which are essential macronutrients used for indiscriminately to boost up crop growth. Excessive uses of fertilizers cause micronutrient imbalance and will cause decrease in productivity of the soil. In addition to this, fertilizers applied in the fields often leach deep into the soil and ultimately contaminate the ground water (Grant and Sheppard, 2008). The nitrates in the nitroginated fertilizer get concentrated in the water and when their concentration exceeds then requirement. It becomes the cause of a serious health hazard for the living beings called blue baby syndrome. This disease affects the infants to the maximum extent causing even death. Excessive use of these types of fertilizers in the agricultural system leads to another problem, which is to water bodies. The contents of fertilizers like nitrogen and phosphorus washed off from the field along with runoff water. This will reach the water bodies though flowing water causes Eutrophication. This is one of the major adverse effects on the environment in many ways. Because of eutrophication, the algal species grow very fast by rapidly using up the nutrients and are toxically affects the food chain. In continuation, these algal species quickly complete their life cycle. Lot of dead matter starts getting decomposed by producing other organic waste. The decomposition process consumes oxygen and very soon the water gets depleted of dissolved oxygen. This kind of processes effects aquatic life in shorter period and it will not be solvable for long time. Due to these serious effects the usage of the fertilizers may be reduced by use of natural carbohydrate base fertilizer in a closed assembly (Serpil, 2012; Megharaj *et al.*, 2011).

Effects of Pesticides

Usually pesticides are poisonous chemical based liquid used as biological agent, antimicrobial, disinfectant used against any pest. Herbicides and insecticides are commonly used to kill the some plants as well as insects. Insecticides are used in agriculture, medicine, industry and the household and are in agricultural activity. A pesticide poisoning occurs when chemicals intended to control a pest affect non-target organisms such as humans, wildlife, or bees. Pesticide use raises a number of environmental concerns. Over 98 per cent of sprayed insecticides and 95 per cent of herbicides reach a destination other than their target species, including non-target species, air, water and soil. Pesticide drift occurs when pesticides suspended in the air as particles are carried by wind to other areas, potentially contaminating them. Pesticides are one of the causes of water pollution, and some pesticides are persistent organic pollutants and contribute to soil contamination. In addition, pesticide use reduces biodiversity, reduces nitrogen fixation, contributes to decline, destroys habitat and threatens endangered species. Agricultural pests are insects, nematodes, bacterial and viral diseases, weeds and vertebrates (Aqsa *et al.*, 2014; Aneja *et al.*, 2009).

Effect by Water Logging

Over irrigation of croplands by farmers for good growth of their crop usually leads to water logging. Inadequate drainage causes excess water to accumulate underground and gradually forms a continuous column with the water table. Under water-logged conditions, pore spaces in the soil get fully drenched with water and the soil-air gets depleted. The water table rises while the roots of plants do not get adequate air for respiration. Mechanical strength of the soil declines, the crop plants get lodged and crop yield falls (Bowonder *et al.*, 1986).

Effect by Salinity

Salinity refers to increased concentration of soluble salts in the soil. It results due to intensive agricultural practices. Due to poor drainage of irrigation and flood waters, the dissolved ill o in these waters accumulates on the soil surface. In arid areas with low rainfall, poor drainage and high temperatures, water evaporates quickly from the soil leaving behind the salts in high concentration. Excess of these salts (mainly carbonates, chlorides and sulphates of sodium and traces of calcium and magnesium) from a crust on the soil surface and are injurious to the survival of plants. The water absorption process of the plant is severely affected (Caro *et al.*, 2012).

Effect by Crop Crushing

Usually formers will cut the plant products on grain part to separate the products from other unrequited part of the plants. Bulk cutting the plants gives lot of waste materials and in under dry form. Separation of the required products from unwanted materials is done by heavy air flow. The weighted required product settle down by leaving light waited unwanted materials. This kind of materials enter in to the environment caused the disturbance in the content of the environment. The small agricultural dust particles float in the environment, which may give health hazard for every leaving being on the earth (Krassi *et al.,* 2019).

Effect of Anthropogenic Soiling

Anthropogenic soiling can be done by mixing of different kind soils for better cropping. Each soil and muddy materials of its own content and may not be proper soiling. This kind of improper soiling may give some unwanted gross like small plants and is considered as waste byproducts. The process of separation of this waste product is considered as difficult and additional labor based activity. It also hinders the growth of required plant products. The formation of unwanted microproducts after drying and it may again mixing with soil or reaches to the troposphere as a pollutant (Jat *et al.,* 2020). The detailed impact of environment by agricultural activity is shown the Figure 12.1.

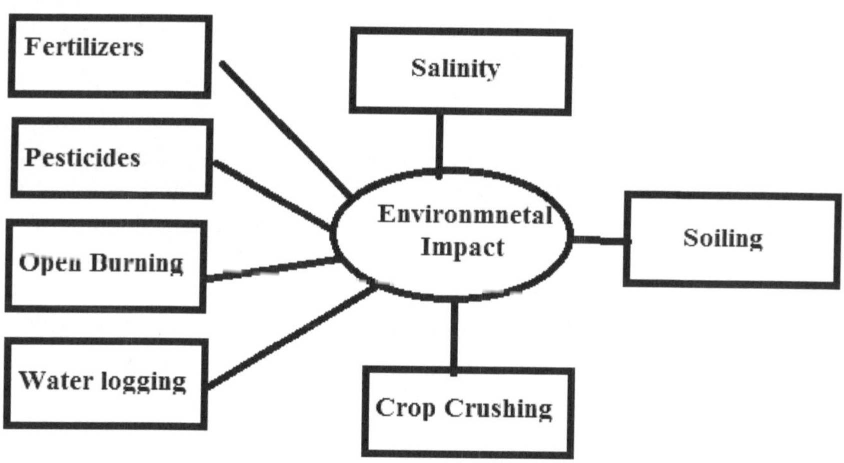

Figure 12.1: Impact of Agriculture on Environment.

Effects of Housing Activities on Environment

Housing characteristics includes home decorations; pet keeping and other environmental factors have significant impact on the health of the residents. The poor housing can have profound, directly measurable effects on physical and mental health of the residents. The air-tight sealing of modern buildings has helped in reducing spiraling energy costs but it has contributed significantly to the health problems arisen due to indoor air pollution. Several building materials continue to affect the indoor air quality. These materials used as solvents, finishes and cleansers for maintenance and protection of building materials can cause 'sick building syndrome'. Production of plastics is also associated with generation of the greenhouse gas namely carbon-dioxide, volatile organic compounds and polyvinyl chloride, which are harmful because of global warming potential of CO_2 and health related problems of the latter two. Manufacturing of metals from their ores has several environmental impacts (Terry *et al.,* 2006). In the recycling of metals harmful chemicals dioxins are produced, which are carcinogenic (cancer causing) in nature. Some insulating materials are made from non-renewable petroleum resources, while some use chlorofluorocarbons. During demolition, their safe recovery is difficult. Release of CFCs in the atmosphere would enhance the global warming problem. Asbestos, which has been quite in use in buildings is now known to be very harmful for our health and not recommended now. The indoor air pollution is a major source of public exposure to air pollutants having potential to cause chronic health problems. It is observed that, over the last two decades have shown measurable levels of some carcinogens in modern offices and homes. The sources of fumes, vapours or gases of indoor air pollutants or chemicals are extremely varied. They may be associated with the building materials used in the construction, the furnishings put in them, types of equipment used for heating and cooling, natural processes that allows gases to seep into buildings, manufacturing process of products like plywood or laminate cabinet doors and surfaces. The potential sources of indoor air pollutants or chemicals are incredibly varied. They can arise from both human activity and natural process. Some of the key indoor air pollutants along with their sources are stated as under: The sources of formaldehyde in modem buildings include building materials, smoking, household products, and the use of un-vented, fuel-burning appliances (Maræ *et al.,* 2018). The most significant sources of formaldehyde are likely to be pressed wood products made using adhesives that contain urea-formaldehyde resins. Benzene is a solvent used in petrol, ink, oil, paint, plastic and rubber. Trichloroethylene is used in metal degreasers, dry cleaning solvents, inks, paints, lacquers, varnishes and adhesives. Air conditioning equipments, which harbor the disease-causing bacteria in air ducts and also in filters. It transports these pathogens through a building as a bacterial aerosol when heating or cooling units are in use. However, spread of the disease is not limited to this pathway. One epidemic occurred in a hospital as a result of contamination from an adjacent construction site. Out-dated heating and cooling systems with inadequate filtering systems can aggravate the problem. Then there are the molds, mildews and other household causes of air borne allergens, dander or dust in the home. Add to this the toxic components of fiberglass insulation, tightly sealed windows and doors, pets, and even human dander, and indoor air quality can easily be quite dramatically impacted (Vornanen-Winqvist *et al.,* 2018; Leung, 2015). Some varieties of asbestos, used as an insulating material and fireproofing material in homes, schools and offices, are known to cause a particular type of lung cancer. Even our kitchen trash-bin contributes a huge volume of illness-causing bacteria and unpleasant odour to indoor air. The Figure 12.2 represents some of the housing activities responsible for impact on environment.

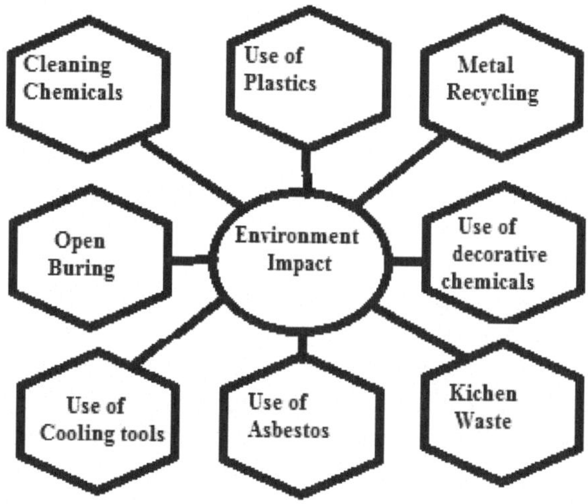

Figure 12.2: Sources of Housing Activities for Environmental Impact.

Mining Effects

Mining is the extraction of minerals and metals from earth. Mining metals like Manganese, tantalum, copper, tin, nickel, bauxite, iron, gold, silver, and diamonds are extracted from the earth crest. These mined metals find many applications such as in cell phones, pagers, and lap-tops, to make pipes, cookware, *etc.* (Kumar and Kumar, 2014) . Gold and silver are extracted by mining process and are high demand in jewellery materials. The mineral resources lead the metallic and nonmetallic products and further ferrous and non-ferrous leads from metallic and nonmetallic sources. The types of mineral resources are given in the Figure 12.3.

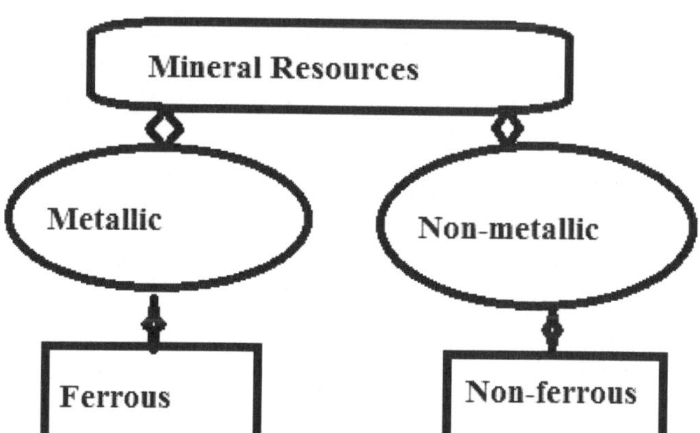

Figure 12.3: Mineral Resources.

The mining procedure and other parameters impacts lot on environment (Fernando, 2017). The environmental impact varies with the stage of development of the resource, *viz.*, exploration, mining, processing and dereliction stages. Large scale mining usually involves a company with many employees. The company mines at one or two large sites and usually stays until the mineral or metal is completely excavated. Small scale mining usually involves a small group of labors. They travel together and look for sites which they think will yield gold or another valuable metal or mineral. It is believed that, small scale mining is more harmful to the environment and causes more social problems than large scale mining. It is equally devastating to the environment which leads climate change. Land dredging and river dredging are considered as two small scale mining in which land dredging involves miners using a generator to dig a large hole in the ground. They use a high pressure hose to expose the metal-bearing layer of sand and clay. The metal bearing slurry is pumped into a sluice box, which collects metal (say gold) particles, while mine tailings flow into either an abandoned mining pit or adjacent forest. During rainy season, the mining pits fill with water and become stagnant water pools. These filled pits create developments of water pollution and also produce water-born insects (Mondal *et al.*, 2014). This kind of pollution able activity may form the water born diseases which effects especially on human beings. The Figure 12.4 shows source, mining process and waste formation after the complete process.

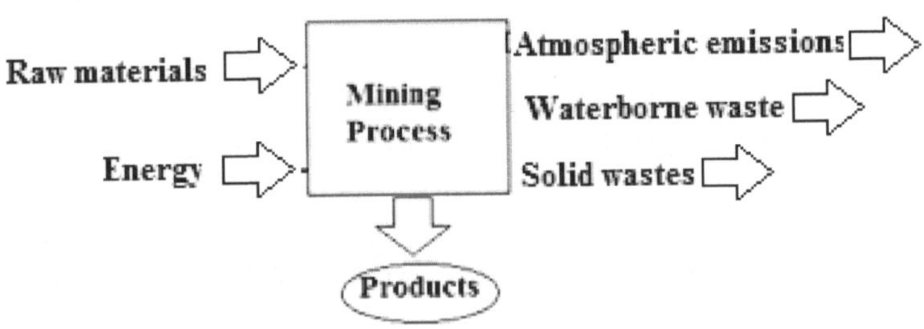

Figure 12.4: Mining Process and Wastes as Byproducts.

After the mining process, the remaining gravel, mud, and rocks go back into the river, but in a different location than where it was originally suctioned. This may be one of the response of the water pollution. Displaced gravel and mud disrupt the natural flow of the river in which the living organisms of the river may die due poisoning the system. Mining is generally considered as destructive process and causes deforestation which impacts lot on the environment.

In order to cluster the extractions, they use chemicals such as cyanide, mercury, or methyl mercury. These chemicals go through pipes and are often discharged into rivers, streams and oceans. This pollution affects all water bodies and ultimately the effect on public. Environmental effect by de-vegetation and defacing of landscape is usual process by mining activity. In addition to this, subsidence of land, ground water contamination, surface water pollution, air pollution, occupational health hazard and climate change are other effects by mining activity. The reduction of such mining process or improvement of the extraction process may be very essential requirement for the healthy environment. This can be done by plantation, make some continuous water flow instead of stalling and avoid harmful chemicals in the process.

Health Issues by Pollution

Due to modernization many essential processes and essential requirements causes environmental pollution. Different essential processes produce many harmful products to the environment and are impacted badly on living beings. Major health issues especially in human beings are responsible by such harmful products. Different pollutants like water and air pollutants effects adversely and leads to different diseases. Some of the important pollution able process and its health issues are given in the Table 12.1.

Table 12.1: Essential Polluting Process and Health Issues

Sl. No.	*Essential Process*	*Health Issues*	*Wastes Forms*
1	Combustion	Dizziness, headache, confusion, nausea/emesis, fatigue, tachycardia, eye and upper respiratory tract irritation, wheezing/bronchial constriction, persistent cough, elevated blood carboxy hemoglobin levels and increased frequency of angina in persons with coronary heart disease.	Solids, gases, microparticles
2	Animal and dust	Recognized infectious disease, exacerbation of asthma, rhinitis, conjunctiva inflammation, recurrent fever, malaise, dyspnea, chest tightness and cough.	Solid, liquid and gaseous wastes
3	Volatile/ organic compounds	Conjunctiva irritation, nose and throat discomfort, headache, allergic skin reaction, dyspnea, declines in serum cholinesterase levels, nausea, emesis, epistaxis (formaldehyde), fatigue and dizziness.	Gases
4	Lead Vapors	Gastrointestinal discomfort, constipation, anorexia, nausea, fatigue, weakness, personality changes, headache, hearing loss, tremor and lack of coordination; whereas in case of infants and small children, the symptoms of lead poisoning are -irritability, abdominal pain, ataxia, seizures/ loss of consciousness and chronic learning deficits, hyperactivity and reduced attention span.	gases
5	Mercury vapours	Muscle cramps or tremors, headache, tachycardia, intermittent fever, acrodynia, personality change and neurological dysfunction	Gases
6	Nuclear energy programes	Anemia, Leukemia and cancer, Embryo defect	Solid and gases

Sl. No.	Essential Process	Health Issues	Wastes Forms
7	Open burning	Respiratory disorder, suffocation, headache, Giddiness, Inability to distinguish, Intelligence of the children, restlessness Asthma	Gases, solids and micro particles
8	Chemical process	Allergic Asthma, Stress on heart, Chronic bronchitis, Acid Rain	Oxide gases

Conclusions

The chemical hazards of the environmental are a function of its environmental persistence, toxicity, and bioaccumulative potential and are more hazardous. Most of the wastes are considered hazardous due to development of toxic environment. Some heavy metals and metalloids from the various sources include Cr, Ni, Cu, Zn, Cd, Pb, Hg, and As are treated as poisonous for the environment and is important implications for wildlife and human health. It is necessary to monitor the concentrations of pollutants in the environmental for healthy environment. A detailed study related to the chemistry of the environmental and hazardous nature of the environment should be taken to minimize the impact of unwanted products on human health and the environment. The public should be educated about the harmful effects of pollutants or wastes on human health and the environment. In addition to this, scientific research on environmental assessment of wastes should be encouraged and promoted in the form some research projects to the researchers with appropriate research funds. An effective support from governmental or non-governmental organization is very essentially required for improvement of environmental conditions as for as human health is concerned.

Acknowledgments

Author Prof. A. Venkataraman, Professor, Department of Chemistry, Gulbarga University, Kalaburagi, Karnataka, India for useful discussion in drafting the MS.

REFERENCES

Ahmad, K. M. and G. A. Mujahid, Environmental Pollution: Its Effects on Life and Its Remedies (July 1, 2011). Researcher World: Journal of Arts, Science and Commerce, 2(2)(2011)276-285.

Ali, H. and E. Khan, "Environmental chemistry in the twenty-first century, " Environmental Chemistry Letters, 15(2)(2017)329–346.

Aqsa Abbasi, Ayesha Sajid, Namra Haq, Sammia Rahman, Zujaja-tul Misbah, Gul Sanober, Muhammad Ashraf, Alvina Gul Kazi. Agricultural Pollution: An Emerging Issue. (2014) 347-387

Bowonder, B., K. V. Ramana, and R. Rajagopal, "Waterlogging in irrigation projects, " Sadhana, 9(3)(1986) 177-190.

Cao G, Wang XG, Liu Y, Luo WB. Effect of water logging stress on cotton leaf area index and yield. Procedia Engineering. 2012; 28: 202–209.

Chen, C., Wang, Y., Qian, Y., Zhao, X., and Wang, Q. The synergistic toxicity of the multiple chemical mixtures: implications for risk assessment in the terrestrial environment. Environ. Int. 77(2015)95–105

Díaz, G., Macià, H., Valero, V. An Intelligent Transportation System to control air pollution and road traffic in cities integrating CEP and Colored Petri Nets. Neural Comput and Applic, 32(2020) 405-426.

Fernando P. Carvalho, Mining industry and sustainable development: time for change, Food and energy security, 6(2)(2017)61-77

Grant, C. A. and S. C. Sheppard, "Fertilizer impacts on cadmium availability in agricultural soils and crops, " Human and Ecological Risk Assessment: An International Journal, 14(2)(2008)210–228.

Heys, K. A., Shore, R. F., Pereira, M. G., Jones, K. C., and Martin, F. L. Risk assessment of environmental mixture effects. RSC Adv. 6(2016)47844-47857

Investigation of Effect of Chemical Fertilizers on Environment, Serpil Savci, APCBEE Procedia 1 (2012) 287-292

Jat, H.S., Madhu Choudhary, Ashim Datta, A.K.Yadav, M.D.Meena, Ritu Devi, M.K.Gathala, M.L.Jat, A.Mc Donald, P.C.Sharma, Temporal changes in soil microbial properties and nutrient dynamics under climate smart agriculture practices, Soil and Tillage Research, 199(2020) 104595.

Jeffrey, J., P. Peirce, A. Vesilind, R. Weiner, Environmental pollution and control, Elsevier Publication, 1997

Kansal, B.D. and Singh, J., Influence of the municipal wastewater and soil properties on the accumulation of heavy metals in plants, Int. Conf. Heavy Metals in the Environment, Hadelberg, (1983) 413-416

Kiran B. Chhokar, Mamata Pandya and Meena Raghunathan, Understanding Environment SAGE Publications India Pvt. Ltd, 2013.

Kloke, A., Reuse of sludges and treated wastewater in agriculture, Water Sci. Tech., 14(1982)61-72.

Krassi Rumchev, Suzanne Gilbey, Ryan Mead-Hunter, Linda Selvey, Kevin Netto and Ben Mullins Agricultural Dust Exposures and Health and Safety Practices among Western Australian Wheatbelt Farmers during Harvest, Int. J. Environ Res Public Health. 16(24) (2019)5009.

Kumar P, Kumar N; Environmental Impact Assessment (EIA) Study of Non-Metal Mines: A Critical Review, International Journal of Engineering and Technical Research (IJETR), 2014; 2(5): 324-326.

Leung D.Y.C. Outdoor-indoor air pollution in urban environment: Challenges and opportunity. Front. Environ. Sci. 2015;2: 69.

Maræ M., Œmie³owska M., Namieœnik J., Zabiega³a B. Indoor air quality of everyday use spaces dedicated to specific purposes–A review. Environ. Sci. Pollut. Res. 2018;25: 2065–2082.

Mark L, Brusseau, Ian L. Pepper and Charles P Gerba Environmental and pollution Science, Elsevier publication, 2019.

Megharaj, M., Ramakrishnan, B., Venkateswarlu, K., Sethunathan, N., and Naidu, R. Bioremediation approaches for organic pollutants: A critical perspective. Environ. Int. 37(2011)1362–1375.

Mondal, S., Bandayopahyay, J. and Chakarvarty, D. Scientific Investigation of the Environmental Impact of Mines Using Geospatial Techniques over a small part of Keonjhar District of Orissa. International Journal of Scientific and Research Publication. 4(2014)1-8.

Terry Norgate, Sharif Jahanshahi, William Rankin, Assessing the environmental Impact of metal production processes, Journal of Cleaner Production 15(8) (2006): 838-848

Viney P. Aneja, William H. Schlesinger and Jan Willem Erisman, Effects of Agriculture upon the Air Quality and Climate: Research, Policy, and Regulations, Environ. Sci. Technol., 43(12)(2009)4234–4240

Vornanen-Winqvist C., Salonen H., Järvi K., Andersson M.A., Mikkola R., Marik T., Kredics L., Kurnitski J. Effects of ventilation improvement on measured and perceived indoor air quality in a school building with a hybrid ventilation system. *Int. J. Environ.* Res. Public Health. 2018;15: 1414.

Chapter 13

The Intersectionality of Cultural Diversity of Indigenous People and Local Biodiversity: Perspective of Biodiversity Management

Mrinal Mukherjee

Assistant Professor, Department of Teacher Education,
The West Bengal University of Teachers' Training,
Education Planning and Administration (WBUTTEPA)
25/2 and 25/3 Ballygunge Circular Road, Kolkata – 700 019, West Bengal
e-mail: dr.mmrinal@gmail.com

The world is witnessing the rapid loss of biodiversity than ever before. A consensus about 'Biocultural Diversity' emerged as functional dynamism of inter-relationship and inter-dependence of biodiversity and cultural diversity of indigenous people. 'All societies argue for policies and practices that help sustain their cultures and systems of knowledge' which is instrumental in protection and management of biodiversity. UN Permanent Forum on Indigenous Issues, (2007) define 'Traditional Knowledge System' as a collective form of experiential knowledge, practices and representations evolved and maintained by peoples with their extended histories of inherent interaction with the natural environment, which encompasses expressions of values based cultural beliefs, custom and community laws, accumulated experience regarding land and ecosystem management. CBD, Rio Summit (1992) emphatically stressed the importance of the knowledge, practices of indigenous and local communities for the conservation and sustainable use of biodiversity. Biodiversity loss is pressing challenges to the ecological services of nature that nations face collectively. India is extremely diverse in terms of ecosystem and aborigines. Therefore, it is needed to document the effective participation of traditional knowledge holders that is aboriginal local communities in sustenance of biological diversity and what are the common threats they are facing. In this article, the researcher has explore the existing literature and local biodiversity and the role of Traditional Knowledge System of aboriginals in the sustenance of all life form of nature.

Keywords: *Biodiversity, Indigenous people, Traditional knowledge system, Intersectionality.*

Introduction

The COVID-19 pandemic has further proved the intricate relationship between human and planetary health and it has been estimated that nearly 75 per cent of new infections of human being are zoonotic in nature. It further indicates that any environmental manipulation caused by human activity not only keep ecosystem under stress or consumption of any wild organism either illegally trapped and tirade or cultivated in stressed condition may enhance the scope of animal to human infection of deadly pathogens. World Bank (2020). IPCC (2019): Climate Change and Land report claimed that 75 per cent of Land on the planet and 66 per cent of oceans have severely manipulated by anthropogenic intervention and encroachment. More or less one million faunal and floral species are facing threat of extinction due their natural habitat destruction. In such context from the perspective of conservation and management of biodiversity, year-2021 is a crucial one as it will be beginning of UN Decade on Ecosystem restoration (2021 – 2030) with an intension to massive scale up the restoration of the degraded ecosystem to mitigate climate crisis with focus of biodiversity protection and the Convention on Biological Diversity (CBD) supposed to build up a new framework that would be able to set prioritiesmore specifically for protecting and judicious use and community mutual sharing of the benefit of resources of biological diversity, in place of previous one set a decade ago in Aichi, Japan. The efforts to evolve effective economy management modeling will assess the financial impacts of the decline of ecosystem services in order to select the most instrumental policy that may bring solution for a resilient economy in a healthier resilient planet.

It is high time to estimate the anthropogenic impact on rate of species extinction at global scale and in general considerations the species extinction are 100–1000 times pre human or background extinction levels. The fossil based evidence, molecular phylogenies and statistical modeling all the evidences and data based modeling assumed that the typical rates of background extinction may be closer to 0.1 species per million species per year. Thus, present rate of extinction are 1,000 times higher than natural background frequency of extinction and probability of future rates of extinction may be 10,000 times higher. So the gravity of crisis of biodiversity loss is more than we can imagine and therefore the issue of management and protection of biodiversity is utmost important.

Every living form on earth is part of biodiversity. But more technically speaking there are three main levels that make up biodiversity, these are genes, species and ecosystems. And these category or levels can further also be described in terms of their attributes:

Components: the identity and variety of the genes, species and ecosystems

Patterns: the unique pattern of spatial distribution of genes, species, habitat in different scale range of landforms. The uniqueness of distribution patterns is important attributes.

Processes: evolutionary processes engaging ecological interactions of genes, species and their immediate ecosystems.

The Figure 13.1 adapted from Peck (1998). 'Attributes of the biodiversity hierarchy' explain the attributes of the hierarchy of biological diversity. The components, patterns and processes collectively develop and control the structure and functions of the ecosystems, including factors that maintain the ecosystems and the human value.

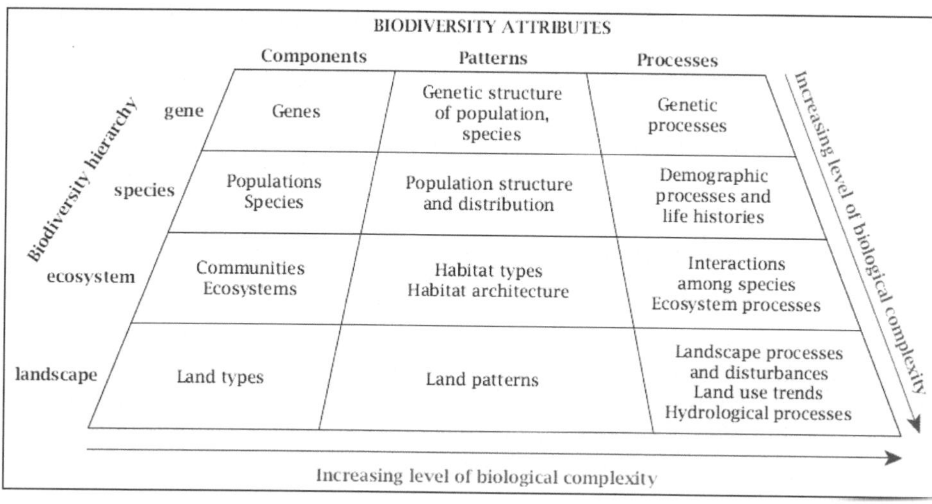

Figure 13.1: Attributes of the Biodiversity Hierarchy (Peck, 1998).

The above model indicate how the diversity hierarchy components are interacting with diversity attributes components and it is indicating that human engaging process are impacting the intensity of biological diversity if the nature allowed to operate in normal course. The role of demographic settlement is crucial and hence the cultural aspects of each and every unique ethnic group may play instrumental role in sustenance of biological diversity.

The relationships between biodiversity and Cultural practice, have been the subject of research over recent decades (Agnoletti, 2006) while human societies have interacted with nature for thousands of generations (Heckenberger *et al.*, 2007) and remain harmonious relationship protecting each other. This is reflected in all cultures by long histories of authorities and rules to protect and preserve natural places in the form of sacred sites, nature reserves and as community conserved areas. The way of life of human being has influenced the other forms of life in nature and being influenced by other creatures. So the culture of human societies co-evolved in many ways with the other form of life in nature. Concept of 'cultures' exist in many different contexts, majority of cultural ideas and practices of aboriginal and indigenous people are beneficial for the nature conservation and thereby created conducive climate for speciation but all cultural ideas and expression are not necessarily nature friendly. Cultural diversity is above all a fact: there exists a wide range of distinct cultures. 'Cultural diversity, beyond the mere fact of its existence, has aesthetic, moral and instrumental value as the expression of human creativity, the embodiment of human strivings and the sum of humanity's collective experience' (UNESCO, 2009). In earlier context interlink of biological and cultural diversity have been considered as unrelated identity and phenomena (Maffi 1998;Berkes 2001; Higgins-Zogib *et al.*, 2010). In most recent time there is an emerging consensus that the variety of form of living organism and human beliefs, values, worldviews and cosmologies are inter-related (Berkes *et al.*, 2000; Maffi 2001; Harmon 2002). There is substantial research evidences of the interconnectedness of human cultural perspective with conservation of biodiversity. Cultural diversity has become a key concern, as a resource to be preserved amid accelerating globalization leading to the homogenization of cultural models and thereby it will ensure biodiversity with different unique practice of life. Cultures are not static entities; they overlap and interact and thus culture is evolving and such evolution of culture and biodiversity has been intricately dependent with each other.

The world is not witnessing the rapid loss of biodiversity, but it has also been witnessing the displacement and extinction of unique ethnic human settlement. Loss of any unique ethnic group is the loss of a unique set of indigenous knowledge. Even forcefully altered practice of life of indigenous people is a steady loss of cultural knowledge of nature and in particular knowledge of management of landforms and the subsequent changes in stakeholder perceptions and 'sense of place. As it is evident that the sustenance of the biological diversity is associated with life and culture of the traditional knowledge of indigenous people, therefore, It is time to recognise biocultural assets within eco cultural landscapes (Merçon *et al.*, 2019; Rotherham, 2015) and the significance of acknowledging 'intangible cultural heritage' (Rotherham, 2007).Cultural heritage includes living expressions inherited from ancestors. This unique experience of living are transmitted to the next generation including oral traditions, performing arts, social practices, rituals, festive events, knowledge and practices concerning nature and the universe (UNESCO, 2018)

The issue of interconnectedness of cultural diversity and biodiversity was a big issue since 1980s or even before than that. The Declaration of Belém (1988), developed in the First International Congress of Ethnobiology, took place in Belém, Brazil attended by indigenous and traditional peoples (It is also highlighted in Article 8(j) Convention on Biological Diversity (CBD), 2021, as 'indigenous and local communities embodying traditional lifestyles') express concern of massive reduction in diversity of both nature and culture worldwide. How indigenous and aboriginal peoples perceived mother nature used and managed earth resources were focus of discussion and strategies were formulated to promote, preservation and strengthening these communities and of their traditional knowledge of cultural practices. The declaration, noted: 'That there is an inextricable link between cultural and biological diversity'. And local culture was in priority. As a consequence of local regional and international efforts, later on in 1992 in Rio Conference the Convention on Biological Diversity (CBD) appeared as a major global functional forum to discuss nature (biodiversity) conservation (United Nations, 1992). In Article 8(j) of the convention text which ensure each country/Contracting Party '....subject to its national legislation, respect, preserve and maintain knowledge, innovations and practices of indigenous and local communities embodying traditional lifestyles relevant for the conservation and sustainable use of biological diversity and promote their wider application with the approval and involvement of the holders of such knowledge, innovations and practices and encourage the equitable sharing of the benefits arising from the utilization of such knowledge, innovations and practices'. In this backdrop it is need to relook the entire episode of dynamics of interface of biological diversity and cultural diversity of indigenous and aboriginal people.

Intersection of Cultural Diversity and Biological Diversity

Biodiversity is conceptualized as the variation of life at the layers of genes, species and ecosystems (United Nations: CBD 1992). The conservation of biodiversity is related with 'ecosystem health' which denotes the potential of natural habitat to maintain their function by controlling unwanted intervening factors. Such resilience is an compulsive criterion to sustainable live practice, human health and relevant social objectives, as drafted in the MDG (Rapport, 2006; Stolton *et al.,* 2008, asserted that biological diversity is the key component to resilient ecosystem, as it protects from environmental shocks and stresses by acting as an absorptive barrier. The traditional indigenous pro-environmental cultural practices is a core components of ecosystem health that ensure sustenance of biological diversity.

A culture is dynamic and complex process having potential to change rapidly, emerging as well as undergoes extinction in both pre-industrialized and industrialized era. In a likewise manner biodiversity pinpoints the resilience of natural systems and cultural diversity has the capacity to increase the resilience of social systems (Gunderson and Holling 2002; Harmon 2002). Human cultural diversity is "the system of shared symbols, behaviours, beliefs, values, norms, artefacts and institutions that the members of a society use to cope with their world and with one another, and that are transmitted from generation to generation through learning"(Brey 2007). A culture is not important for mere utilitarian function as Smith (2001) stated "a form of co-evolution between cultural information and the social and natural environment".Such definition help us to interpret 'culture' as a complex dynamic and intrinsic system of interlinked components that attributes to an individual's identity in the authentic context of its relationships with immediate environment. Different form cultural practices interact with nature in different manner and engage in localized unique relationships with their immediate environments (Berkes 2012).

Nature and culture converge at multiple layer including values, beliefs, norms, livelihoods, knowledge and languages. The natural environment is the petri dish for harboring the cultural processes, allow to inculcate belief systems and subsequently, landscapes form a diverse cultural archive of human endeavors (Schaaf and Lee 2006; Berkes 2012). These dynamic inter-playcreate natural feedback mechanism, and any shift in one system mostly leads to a change in the other (Maffi and Woodley 2007; Pretty 2012). There are examples in nature where the innate biodiversity are extremely crucial for cultural diversity, indicated by the density of ethnic groups and linguistic diversity (Skutnabb-Kangas *et al.,* 2003; Sutherland 2003). Research findings identified four major bridges linking nature with culture; beliefs and world views, livelihoods and practices, knowledge bases and languages, and norms and institutions. A level of analysis in traditional knowledge and management system is explained after Berkes *et al.* (2000).

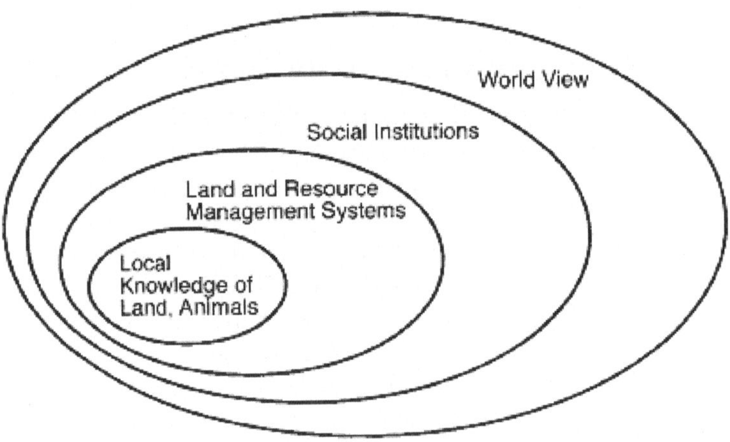

Figure 13.2: Rediscovery of Traditional Ecological Knowledge as Adaptive Management (Source: Berkes, 2000).

A consensus about 'Biocultural Diversity' emerged as functional dynamism of inter-relationship and inter-dependence while CBD COP (2018) had accepted the following definitions: 'Biocultural diversity is considered as biological diversity and cultural diversity and the links between them' and 'Biocultural heritage' reflects the holistic approach of many indigenous peoples and local communities. Thus the concept of custodial and intergenerational attributes is acknowledged. Human culture and nature are mutually constituting and evolving, hence it established that biodiversity and cultural diversity are interlinked and the interconnectedness have structured over long period of mutual adaptation. Thus life practices of indigenous people have impacted the richness of biodiversity and are mutually influenced by the exhibiting location based innate biological diversity. Biocultural diversity incorporates ethno-biodiversity. Diaz *et al.* (2015), in the context of the IPBES Conceptual Framework proposed-

1. The inclusion of bio-cultural matters in differing worldviews was considered in the framework of the 'The Economics of Ecosystems and Biodiversity' (TEEB, 2010).
2. Sukhdev, Wittmer, and Miller (2014) recognize that values are a product of different worldviews and perceptions on the relationship of humans and nature. They treat them as legitimate and valid in their respective socio-cultural contexts.
3. TEEB (2010) argues that neither is incorrect nor invalid in their respective socio cultural contexts, as values are always derived from worldviews and perceptions. Because of this multi dimensional and socio cultural embeddedness of "value", any exercise of valuation is purely a reflection of how certain people perceive their natural environment, and their relationship to it, at a certain point in time.

The natural evolving network of interconnections of the human culture and biological diversity are hampered and facing challenges from the same 'drivers and threats' and as result severe biodiversity loss has been actualized in last couple of decades and such huge loss has been triggered by some common drivers (Pretty *et al.*, 2007, Pilgrim *et al.*, 2008; Pretty *et al.*, 2009).

Such threats are so called standardized paradigms of development in urbanization, industrialization, healthcare system, education reforms *etc.* in accelerated globalization. More clearly speaking for example imposition of culturally inappropriate health care and education on the indigenous people has negatively impacted their way of life which leads to language erosion, hampering the inter-generational knowledge transfer and dismantling the local knowledge base and mostly alarmingly the aboriginal indigenous people started to spent less time in exposing and interacting with nature under the cognitive apprenticeship of the family members and the community leaders. (Pilgrim *et al.*, 2007). Displacement of local people from their land, privatization of land which were natural habitat of indigenous people, policy of exclusion and massive urban migration altered the course of life and thus natural resource management system was became fragile and caused biodiversity loss in a high degree. Alarming crisis is that the erosion and de-escalation of place based culture was the immediate result of physical separation of such indigenous people from the land and their beliefs and world view. Every part of their life practice like stories, narratives, festivities and rituals lose their meaning and merit when placed out of context leading to cultural catastrophe (Samson, 2003).

The mechanistic imposition of so called globalization impacted food consumption behavior lead to monoculture instead of farming of wild variety thus biodiversity facing a great threat. Such change of practice of agriculture and livelihood leads to the loss of ecological knowledge including skills and practices developed standardized and fine-tuned locally in long range of time (Samson and Pretty 2006). Such commodification of products and change of livelihood in massive scale delinked the indigenous people from their land oriented life practices and local traditional knowledge which in turn causing the depletion of biodiversity (Pillgrim *et al.,* 2008). Environmental destruction by means of multinational controlled unsuitable agricultural practices and conversion of forest for agricultural purpose are major driver for decline of local biodiversity and simultaneously place based anthropological culture became fragile and decline due to socio-political and commercial stress induced by imposition of common paradigm of development. In absence of any frame work of transnational cooperation and subsequent geopolitical instability and unwanted and inappropriate interference in human-nature harmonyat global to local level accelerated the process of erosion. Collective negative impact of such drivers has caused heavy loss of biodiversity which is termed as sixth great extinction period (Pretty 2002, Skutnabb-Kangas, *et al.,* 2003). To mitigate the negative drivers and to protect biodiversity by means of protecting cultural diversity identification of such factors are crucial in biodiversity management. The list of the driver as identified by Pretty *et al.* (2009) are listed in Table 13.1.

Table 13.1: Common Drivers or Threat and Outcomes for Biological and Cultural Diversity

Driver or Threat	*Outcomes for Biological Diversity*	*Outcomes for Cultural Diversity*
New commercial resource use (*e.g.* timber, biofuels, energy)	Loss of habitat and species (*e.g.* tropical forest)	Conflict, loss of livelihoods, erosion cultural identity
Extended commodification of natural resources	Loss of biodiversity due to shift towards unsustainable management practices	Departure from traditional resource use and management strategies
In-migration of new economic actors (*e.g.* settlers, ranchers)	Accelerated loss of habitat or species	Conflict, disease, competition for basis of livelihoods erosion of cultural identity
Pollution of water courses) *e.g.* mercury from dams)	Degradation of wetland and freshwater habitats	Loss of livelihoods, erosion cultural identity
Aspirations for consumer lifestyles worldwide	Increased exploitation of natural resources as economic incentives replace subsistence incentives	Loss of traditional management strategies and local knowledge; loss of biological basis for culture and economy

Driver or Threat	***Outcomes for Biological Diversity***	***Outcomes for Cultural Diversity***
Continuing globalization of food system	Loss of on and between farm habitats and species due to monocultures resulting in loss of agro-biodiversity more isolated protected areas	Loss of traditional diets and knowledge of famine foods
Urbanization and rural-urban migration.	Habitat destruction leading to loss of green spaces and species, ecological footprint effects remote from urban area	Erosion of rural cultures; loss of land on which stories and culture center upon, contributing to loss on identity and understanding
Modernization of healthcare	Reduce pressure on wild species, but departure from traditional medicine may lead of these species being devalued and lost	Shift away from traditional resources, loss of ecological knowledge and practices that have long been central to culture and identity
Spread of formal education including the expansion of dominant belief system	Departure from traditional management techniques and knowledge, natural resources becoming devalued; links to other processes (*e.g.* migration)	Changes in mechanisms of knowledge transmission, devaluation of traditional knowledge, inter-generational conflicts, loss of traditional stories and worldviews
Language erosion and loss	Loss of knowledge and traditional management strategies previously passed on verbally between generation	Reduce communication with elder generations leading to loss of knowledge of cultural heritage, local resources and management techniques
Formalization and privatization of land rights	Loss of biodiversity supported by traditional management systems, reduced communal cohesion and common sense of ownership, leading to lack of local support for conservation	Loss of land that stories and culture centre upon loss of identity and understanding changes in social organization sometimes leading to conflicts
State territorialsation and "nation building"	Shift from traditional to external state management of resources at the expense of local species and habitats dependent upon local management regimes	Cultural hybridization including the loss of knowledge, beliefs and practices where individual territories are no longer recognized for their significance to a particular group of people
Expansion of transport networks, including roads	Loss of biodiversity and habitats where construction occurs, heightened levels of pollution	Cultural erosion due to increased access to and interaction with other cultural groups leading to hybridization or assimilation into the more dominant culture
Assimilation	Decline of biodiversity and habitats dependent upon cultural knowledge, community values and the traditional management systems derived from them	Departure from traditional practices and knowledge, leading to the loss of land based management systems, cultural values and worldviews

Cultural Practice of Indigenous People: Perspective of Conservation of Biological Diversity

Delinking nature from human culture is called 'Cultural Severance' which has been identified as a serious threat in the conservation of both nature and heritage. (Rotherham, 2014). The consequences of cultural severance include dramatic declines in ecological richness. Cocks (2006) define 'bio-cultural diversity' as the link between biodiversity and human diversity. Human cultural diversity and biodiversity conservation impact each other as because biodiversity represents a source of raw material on which the processes of evolution actualized. Loss of diversity enhance the potential chance of destruction and abolition of life forms through inadequacy of resilience to shift in environmental structure. Different cultures perceive and appreciate biodiversity in different manners because of their unique heritage and experience (Posey, 1999). 'As the world's remote biodiverse areas are threatened by increasing incursions for "development",the emergency of developing network with members of the indigenous community to protect their lands will create possibilities to increase over the next decade and into the future (Alcorn, J. 2010).It has been documented from research that cultural diversity of indigenous and aboriginal people can sustain a huge variety of practices of use and the conservation of natural resources.

The World Bank Operational Manual, (1991) have identified the following probable characteristics of Indigenous peoples as: (a) close attachment to ancestral territories and to the natural resources in these areas; (b) self-identification and identification by others as members of a distinct cultural group;(c) an Indigenous language, often different from the national language; (d) presence of customary social and political institutions; and (e) primarily subsistence-oriented production. The settlement of Indigenous peoples are distributed in almost all regions of the earth and use approximately 22 per cent of global land area. Nearly 500 million,includes around 5000 different groups, indigenous peoples are in possession of the greater part of the world's cultural diversity, and have created and speak the major share of the world's almost 7000 languages. Most of such ethnic group of indigenous peoples presently facing the challenges of marginalization, extreme poverty and sever deprivation of other human rights. UNESCO acknowledging the instrumental role of indigenous people in continuing the diversity of the world's cultural and biological landscape and simultaneously intend to capacity building of such ethnic human settlement in mitigating the multiple threats and challenges they are facing. Therefore, life of indigenous people and their cultural practice is instrumental in conserving and sustaining the local biological diversity. While indigenous peoples comprise only 4 or 5 percent of the world's population, they use almost a quarter of the world's land surface and manage 11 percent of its forest andthus they maintain 80 percent of the world's biodiversity in, or adjacent to, 85 percent of the worlds protected areas.

World Conference on Indigenous Peoples (2014) in its thematic paper conceptualise 'Traditional Knowledge' as the knowledge, innovations and practices of indigenous and local communities around the world which is based on experiences gather from their life practices through centuries and transmitted to subsequent generation. Such body of knowledge is collectively owned by the community as culture. It covers form of stories, songs,folklore, proverbs, beliefs, rituals, community laws, local language, and agricultural practices.

"The interrelated nature of biological diversity, traditional knowledge, and language necessitates a comprehensive approach to the conservation of biological diversity. Indigenous languages are treasuries of vast traditional knowledge concerning ecological systems and processes. Indigenous languages hold knowledge about how to protect and sustainably use some of the most vulnerable and biologically diverse ecosystems in the world. Losing linguistic and cultural diversity has been directly linked to losing biological diversity." – message of Dr. Ahmed Djoghlaf, Executive Secretary Convention on Biological Diversity on the occasion of the International Day of the World's Indigenous Peoples, 9 August 2008. (Alcorn, 2010). Thus traditional knowledge which is accumulation of cultural practices is a key factor for biodiversity conservation at local level in unique habitats. The International Council for Science (ICSU) also highlighted know-how, practices and representations evolved by indigenous peoples with associated legacy of interaction with the natural environment.

Indigenous traditional knowledge generally means traditional practices and culture and the knowledge of plants and animals and of their methods of propagation. It includes:

- ☆ Expressions of cultural values
- ☆ Beliefs
- ☆ Rituals and community laws
- ☆ Knowledge regarding land and ecosystem management

Rozzi, *et al.* (2006) reinforce the issues of interconnectedness of cultural knowledge and biodiversity conservation and favored a consensus among conservation practitioners of a need for:

1. Social involvement by scientists;
2. Interdisciplinary approaches that integrate human and biological factors;
3. A focus on local, regional and global levels; and
4. The establishment of international agreements on biodiversity and environmental protection, referenced by UNESCO (2000)

UNESCO:COMEST (2015) in the context of climate change and biodiversity loss, in its report entitled "Ethical Principles for Climate Change: Adaptation and Mitigation" proposed that a more comprehensive extension of these principles and responsibilities articulated for climate change adaptation and for protection of biodiversity be undertaken to take into account climate change mitigation. The expanded list of principles identified in relation to climate change includes:

1. Biological diversity;
2. Cultural diversity;
3. Interdependence of life on Earth;
4. Intellectual and moral solidarity of humankind;
5. Global justice;
6. Resilience;
7. Sustainability (frugality, renewable energy, reforestation, water resources);
8. Precautionary principle;
9. The duty to share scientific knowledge; and
10. Integrity of scientific research.

Globally by means of corporate encroachment, political occupancy, so called development projects and uncontrolled urbanization does not only posing threat to the existence of the indigenous unique population settlements but also posing threat of extinction of the native local biodiversity. Forceful detachment of such population settlement from their natural habitat is decreasing their life expectancy and as a result of cultural disassociation of such population from nature fragility of the local biodiversity. Rewilding is the strategy need of the hour to save the biodiversity and it is only possible if the world learned community can understand the emergency of protecting indigenous people in their natural habitat and to allow their indigenous practice of life. The very core essence of indigenous life is to maintain and conserve the immediate biological diversity.

Table 13.2: The Value System of Indigenous and Non-indigenous People towards Nature

Dimensions	*Indigenous People*	*Non-Indigenous People*
A. Natural Resources		
Land	Related to sacred	Ownership, Secular
Environment	Adapt to	Exploit
B. Other Social Values		
Society	Unified	Diverse
Relationships	Extensive	Limited
Basic Unit	Society	Individual
Reality	Spiritual	Material
Possessions	Share, Use	Acquisitive, Accumulative

A substantial research documents are available on traditional knowledge of indigenous people, and local knowledge relevant for nature care. Policymakers across the globe need to aware about the significance of locally developed place based historical and contemporary discourse of knowledge not only sustain *in vivo* protection and conservation of the biodiversity, but also empower local stakeholders to facilitate their survival along with sustenance of nature (Matsui, 2015). As value system of the local population group matters not only in determining their attitude towards nature but also their caring attitude to mother earth. Edward (1988) formulated this comparative table which reflects the difference of value system of indigenous people towards nature in comparison to the non-indigenous counterpart.

So resilience at local level for protection and conservation of biodiversity largely depend on local practice based knowledge. The connection to the land and climate is important input of resilience for indigenous people as this knowledge are locally tested and fine-tuned and can provide long term resilience in protecting diversity of life forms. So the community based approach in biodiversity management is a reflection of identities, cultures, knowledge systems and social organizations of indigenous people. Any form of encroachment in physical territory and cultural space will not only increase the vulnerability of these people but also the local biodiversity. WWF (2013) in its '*participatory biodiversity assessments*' found community engagement remarkably positive impact on biodiversity conservation.

In the context of life and culture of such indigenous people in relation to ecosystem health, the United Nation University in particular seeks to contribute to:

- ☆ Change mindsets and paradigms about the role of traditional knowledge in our society and in key sectors such as academia, government and business
- ☆ Increasing the recognition and importance of traditional knowledge
- ☆ Developing the application of traditional knowledge in a broad range of contexts (*e.g.* ecosystem management and biotechnology)
- ☆ Developing strategies for the preservation and maintenance of traditional knowledge
- ☆ Facilitating the development of the capacity of indigenous communities to conserve and apply their knowledge in an increasingly globalized economy.

Human cultures by means of livelihoods, cosmologies, worldviews and spiritual beliefs interpret and make positive relationship with nature (Berkes 2008). These cultural ethos formulate and regulate individual actions and also collective actions and thus nature is being managed and protected (Milton 1999; Smith *et al.,* 2007). But unfortunately such dynamic and evolving network of interconnections of cultural and biological diversity are rapidly eroded by the same drivers and threats. As local knowledge bases give rise to socially embedded norms and institutions

(Smith *et al.*, 2007) and thereby these contextual systems of collective action are intimately linked to the land on which they are based (Kothari 2006). They exercise community approved informal institutional frameworks and may practice location specific cultural rules and practices on resource exploration and conducive behaviours for the management of natural resources (Smith *et al.*, 2007).

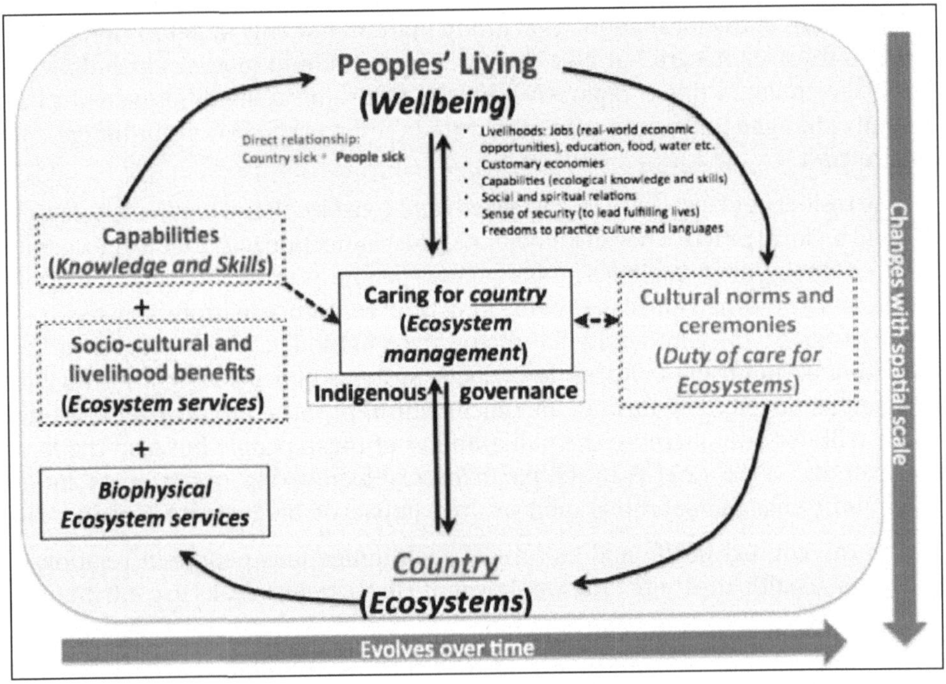

Figure 13.3: An Indigenous Perspective of Living-on Country, Describing Peoples' Connections (non-italic) and the Equivalent Scientific Terms (in italics) (*Source*: Sangha and Russell-Smith, 2017).

The well being of indigenous people can guarantee the well being of nature and thereby biodiversity can be conserved and managed. The concept of 'Ecosystem services framework' that upheld the indigenous peoples' abilities along with well-being perspective could be instrumental input in making policy decision towards the value of natural resources for wellbeing of Indigenous-local peoples'. An integrated framework considering indigenous and local estates as a source of opportunities as advocated by Sangha and Russell-Smith, (2018) might be pertinent enough to understand that this symbiotic relationship could be crucial in management and protection of biodiversity. Such framework is in tune with the implicit feature of peoples' relationships with nature, as advocated by Costanza *et al.* (2017). Any viable ecosystem framework must encompass the diversity of culture of the indigenous local people's views, values and skills and competencies. People have learned to care for and managed their lands by means of their practice of living with nature involving various rituals, customs, ceremonies and traditions.

Sangha *et al.*, 2018 and through their studies, outlined that: i) peoples' relationships with nature involve both people obtaining benefits or services on one hand while undertaking duty to care for and manage their natural landscapes on the other; ii) peoples' capabilities enable them to connect with their traditional lands, and are critical to be passed on from one generation to the next for managing those lands. Such framework could be an important base line and guiding principle to frame policies at global, regional, national and local level.

As far as Indian context is concerned, the traditional knowledge towards biodiversity management and conservation is unique for each 2753 communities (Joshi *et al.*, 1993) according to their geo-climatic distribution, cultural traditions and practice of farming, food habits and subsistence strategies.

Mishra *et al.* (2010) categorized Indian systems of biodiversity management and conservation as:

- ✰ Religious traditions: temple forests, monastery forests, sanctified and deified trees
- ✰ Traditional tribal traditions: sacred forests, sacred groves and sacred trees
- ✰ Royal traditions: royal hunting preserves, elephant forests, royal gardens *etc.*
- ✰ Livelihood traditions: forests and groves serving as cultural and social space and source of livelihood products and services

Mishra *et al.* (2010) have also list up the natural resources protection and management practice of local indigenous people. The variety of resources management practices include:

- ✰ Collection and management of wood and non-wood forest products
- ✰ Traditional ethics, norms and practices for restraint use of forests, water and other natural resources
- ✰ Traditional practices on protection, production and regeneration of forests.
- ✰ Cultivation of useful trees in cultural landscapes and agroforestry systems
- ✰ Creation and maintenance of traditional water harvesting systems such as tanks along with plantation of the tree groves in the proximity

There are lot of documented case studies of interconnectedness of Indigenous people with biodiversity conservation and management in Indian context as advocated by Mishra *et al.* (2010).Such design of practices ensured sustenance of biological diversity in its natural habitat. For example, in case of Rajasthan there are almost 15 types of resource management practices that not only manage the different kind of soil texture in arid ecosystem but also contributed in biodiversity management. Singh *et al.* (2018) studied ecological, socio-cultural and livelihood indicators, is a location-specific practice of indigenous people of eastern Himalayas and found that integrating the indicators identified by the indigenous people in classifying and adapting strategies management of the local forests and bio resources can provide a framework for the sustainable conservation of forest resources. This could be valuable input for future policy making. Cordero *et al.* (2018) investigated the nature of Toda's socio-ecological culture who are inhabitant of Nilgiri Hills, India and land management system extensively. The study suggests that the Toda community have a crucial role to play in protecting and managing the shola-grassland mosaic. Their stewardship role which is the result of their indigenous culture would help conserve this endangered and highly biodiverse ecosystem. Kaur, L. (2015)tried to explore the traditional knowledge of indigenous communities of Rajasthan, India in conservation and management of environment. Their trans-generational pull of experience has the potential to leads to unique discoveries for modern pharmaceutical research. The researcher has the opinion to protect the rights of tribal people and the biodiversity a community biodiversity register (CBR) of local products and their uses may be framed.

Discussion

The sense of well-being of humanity is the reflection of wellness of Mother Nature. While we are trying to renegotiate the policy and approach of sustainability, the discourse of the intellectual resources should essentially consider the evidenced based account the knowledge of local people. Thus any knowledge which might impact the policy of nature conservation and more specifically biodiversity conservation must be the collective intellectual resources of both disciplines of formal sciences along with indigenous knowledge. Therefore, biodiversity management conservation has occupied the epicenter of the contemporary debates and discussions in ecological anthropology and developmental studies as we are facing sever threat of loss of biological and cultural diversity Mishra *et al.* (2010). We are at starting point of understanding of how and in what magnitude the ecosystem diversity is interlinked to the cultural diversity of local people and there is much more to unearth about this interdependence. It is evident from the experience of recent past that these complex systems are receding (Maffi 2001; Skutnabb-Kangas *et al.,* 2003). Vulnerable, and endangered flora and fauna, fragile habitats, decaying languages and vast knowledge bases are being eroding and declining at rates than "natural" extinction rates as there is no effective assessment and policy instrument (McNeely and Scherr 2001; Pretty 2002). Conserving biodiversity with whole ecosphere along with culture of indigenous people has been providing unique challenges (Dove *et al.,* 2005; Robbins *et al.,* 2006). So any expectation of protecting biodiversity or restructuring lost environments is predicated on a concomitant effort to appreciate and protect cultural diversity (Pretty 2012; Maffi 2001; Nietschmann 1992).

There is no dearth of research finding in the field interlink of biodiversity conservation and cultural diversity of local indigenous people, but still there is lack of sound policy frame work to address the right of such local people so that they can continue to protect, manage and restore the richness of floral and faunal diversity. After an extensive literature review, Mishra *et al.* (2010). has identified the focal areas need to be addressed, as follows: a) inadequacy of economic incentives to conserve biodiversity compels rethinking classical utilitarian approach to resource management; b) an emerging sustainability science will need all stocks of knowledge and institutional innovations to navigate transition towards a sustainable planet; c) rediscovery of traditional ecological knowledge as adaptive management and need to apply human ecological and adaptive strategies for natural resource management offers prospects for scientists to address the problems that beset conservation biologists and restoration ecologists; d) there is an increasing realization that we need innovative ethics and policy to conserve biodiversity and maintain ecosystem functions and that such ethics need not come from the god; rather, society can cultivate them; e) local knowledge systems are disappearing at a rate that may not allow us even to know what value, if any, such systems had; f) a limitation of economic development is that it is pursued without any considerations – in practice – as to its implications on ecosystems.

So affirm conceptual and workable frame work for 'biodiversity conservation by indigenous life practices' of the local people may serve as a 'guiding principle' for adopting national policies that will be effective in conserving resources of life support systems in the mutual interest of nature and human society. Such Policies and programs evolved on "bioglobalism" would be appropriate to social democratic practice of society to promote freedom and justice for all (Singh, 2017).Towards this endeavor it is need to document the indigenous knowledge of natural resource management by engaging community along with the threat perceptions they are facing. Keeping the plural fabric of cultural practices of indigenous people unaltered the best practices of biodiversity management need to be shared among the different local communities. Moreover, apart from farming and implementing policy for the right of indigenous people to practice their nature-friendly cultural rituals and livelihood, it is high time to document and develop educational resources of indigenous knowledge system that could be shared by the different groups of local people in one hand and on the other such knowledge system need to be integrate with the formal knowledge system.

REFERENCES

Agnoletti, M., and Rotherham, I. D. (2015). Landscape and biocultural diversity. *Biodiversity and Conservation, 24*, 3155–3165. https: //doi.org/10.1007/s10531-015-1003-8

Alcorn, J. (2010). Indigenous Peoples and Conservation. MacArthur Foundation Conservation White Paper Series. https: //www.macfound.org/media/files/csd_indigenous_peoples_white_paper.pdf

Berkes, F. (2012). *Sacred ecology*. Routledge.

Berkes, F. 2001. Religious traditions and biodiversity. *Encyclopaedia of Biodiversity* (5) 109-120.

Berkes, F. 2008. *Sacred Ecology*, 2nd *Edition*. New York: Routledge.

Berkes, F., Colding, J., and Folke, C. (2000). Rediscovery of traditional ecological knowledge as adaptive management. *Ecological applications, 10*(5), 1251-1262. http: //www.cc.umanitoba.ca/institutes/natural_resources/canadaresearchchair/EA2000.pdf

Brey, P. (2007). Theorizing the cultural quality of new media. *Techné: Research in Philosophy and Technology, 11*(1), 2-18. https: //doi.org/10.5840/techne20071112

Cocks, M. (2006). Biocultural diversity: Moving beyond the realm of "indigenous" and "local" people. Human Ecology, 34(2), 185–200. https: //doi.org/10.1007/s10745-006-9013-5

Convention on Biological Diversity (CBD). (2021). Article 8(j) - Traditional Knowledge, Innovations and Practices. https: //www.cbd.int/traditional/

Cordero, R. L., Suma, M., Krishnan, S., Bauch, C. T., and Anand, M. (2018). Elements of indigenous socio-ecological knowledge show resilience despite ecosystem changes in the forest-grassland mosaics of the Nilgiri Hills, India. *Palgrave Communications, 4*(1), 1-9. https: //doi.org/10.1057/s41599-018-0157-x

Costanza, R., d'Arge, R., De Groot, R., Farber, S., Grasso, M., Hannon, B., and Van Den Belt, M. (1997). The value of the world's ecosystem services and natural capital. *nature, 387*(6630), 253-260. https: //doi.org/10.1038/387253a0

Declaration of Belém. *(*1988*).* The Declartaion of Belem. *Retrieved from* http: //www.ethnobiology.net/wp-content/uploads/Decl-Belem-Eng-from-Posey.pdf

Díaz, S., Demissew, S., Carabias, J., Joly, C., Lonsdale, M., Ash, N., and Zlatanova, D. (2015). The IPBES Conceptual Framework–connecting nature and people. *Current opinion in environmental sustainability, 14*, 1-16. https: //doi.org/10.1016/j.cosust.2014.11.002

Dove, M. R., Sajise, P., and Doolittle, A. (eds.) (2005). Conserving nature in culture: Case studies from Southeast Asia. New Haven: Yale University, Southeast Asia Studies, Monograph No. 54, 348.

Edwards, W.H., 1988. An Introduction to Aboriginal Societies. Social Science Press, Australia.

Gunderson, L. H. and Holling, C. S. (2002). *Panarchy: understanding transformations in human and natural systems*. Washington, DC: Island Press. https: //islandpress.org/books/panarchy

Harmon, D. 2002. *In light of our differences*. Washington, DC: Smithsonian Institution Press.

Heckenberger, M. J., Christian Russell, J., Toney, J. R., and Schmidt, M. J. (2007). The legacy of cultural landscapes in the Brazilian Amazon: implications for biodiversity. *Philosophical Transactions of the Royal Society B: Biological Sciences, 362*(1478), 197-208. https: //doi.org/10.1098/rstb.2006.1979

Higgins-Zogib, L., Dudley, N., Mallarach, J. M., and Mansourian, S. (2010). Beyond belief: linking faiths and protected areas to support biodiversity conservation. *Argum. Prot. Areas Mult. Benefits Conserv. Use Ed. Sue Stolton Nigel Dudley Lond. Earthscan*, 145-158.

IPCC, 2019: Climate Change and Land: an IPCC special report on climate change, desertification, land degradation, sustainable land management, food security, and greenhouse gas fluxes in terrestrial ecosystems [P.R. Shukla, J. Skea, E. Calvo Buendia, V. Masson-Delmotte, H.-O. Pörtner, D. C. Roberts, P. Zhai, R. Slade, S. Connors, R. van Diemen, M. Ferrat, E. Haughey, S. Luz, S. Neogi, M. Pathak, J. Petzold, J. Portugal Pereira, P. Vyas, E. Huntley, K. Kissick, M. Belkacemi, J. Malley, (eds.)].

Joshi, N. V., Gadgil, M., and Patil, S. (1993). Exploring cultural-diversity of the people of India. *Current Science* 64, 10-17.

Kaur, L. (2015). Traditional Knowledge of Indigenous Communities of Rajasthan and Environment (A Short Review). *Journal of Biodiversity and Ecological Sciences*, 5(1).

Kothari, A. (2006). Community conserved areas: towards ecological and livelihood security. *Parks, 16*(1), 3-13. https: //www.academia.edu/download/30283734/ parks_16_1_forweb.pdf#page=5

Maffi, L. (1998). Language: a resource for nature. *Nature and Resources, 34*(4), 12-21.

Maffi, L. (2000). Language, knowledge and the Enviroment: the interdependence ofcultural and biological diversity. Oxford University Press.

Maffi, L. (ed.). 2001. *On biocultural diversity*. Washington, DC: Smithsonian Institution Press.

Maffi, L., and Woodley, E. (2007). Biodiversity and culture. *UNEP's 4th Global Environment Outlook Report.* Nairobi: UNEP.

Matsui, K. (2015). Introduction to the future of traditional knowledge research. *International Indigenous Policy Journal, 6*(2). https: //doi.org/10.18584/ iipj.2015.6.2.1

McNeely, J. A., and Scherr, S. J. (2001). Common ground, common future: How ecoagriculture can help feed the world and save wild biodiversity.http: // www.agriculturesnetwork.org/library/60928

Merçon, J., Vetter, S., Tengö, M., Cocks, M., Balvanera, P., Rosell, J. A., and Ayala Orozco, B. (2019). From local landscapes to international policy: Contributions of the biocultural paradigm to global sustainability. *Global Sustainability, 2*(e7), 1–11. https: //doi.org/10.1017/sus.2019.4

Milton, K. (1999). Nature is already sacred. *Environmental Values, 8*(4), 437-449. https: //doi.org/10.3197/096327199129341905

Mishra, K. K., Mishra, K. K., and Bondla, D. N. (2010). Cultural dimension of biodiversity conservation among the tribes of the Eastern Ghats. *Indian Anthropologist*, 1-20. https: //www.jstor.org/stable/41920107

Peck, S 1998, *Planning for Biodiversity: Issues and Examples*, Island Press, Washington, DC.

Pilgrim, S., D.J. Smith and J. Pretty. 2007. A cross-regional assessment of the factors affecting ecoliteracy: Implications for policy and practice. *Ecological Applications* 17(6), 1742-1751. https: //pdfs.semanticscholar.org/8a24/8f3e9db2b97cb92586a82d8c6ce8407eb1ad.pdf

Pilgrim, S., L. Cullen, D.J. Smith and J. Pretty. 2008. Ecological knowledge is lost in wealthier communities and countries. *Environmental Scienceand Technology,* 42(4), 1004-1009. https: //pubs.acs.org/doi/pdf/10.1021/es070837v

Posey, D. A. (1999). Cultural and Spiritual Values of Biodiversity. A Complementary Contribution to the Global Biodiversity Assessment. In D.A. Posey (ed.), *Cultural and Spiritual Values of Biodiversity*, 1–19.

Pretty, J. (2002). *Agri-culture*: *Reconnecting people, land and nature*. London: Earthscan.

Pretty, J. (2012). *The earth only endures: On reconnecting with nature and our place in it*. Earthscan.

Pretty, J., Adams, B., Berkes, F., De Athayde, S. F., Dudley, N., Hunn, E., Maffi, L., Milton, K., Rapport, D., Robbins, P, Sterling, E., Stolton, S., Tsing, A., Vintinner, E. and Pilgrim, S. (2009). The intersections of biological diversity and cultural diversity: towards integration. *Conservation and Society, 7*(2), 100-112. https: //www.jstor.org/stable/pdf/26392968.pdf

Pretty, J., Ball, A., Benton, T., Guivant, J., Lee, D. R., Orr, D., and Ward, H. (2007). Introduction to environment and society. *The SAGE handbook of environment and society*, 1-32.

Rapport, D.J. 2006. Sustainability science: An ecohealth perspective. *Sustainability Science* 2: 77-84. https: //doi.org/10.1007/s11625-006-0016-3

Robbins, P., McSweeney, K., Waite, T., and Rice, J. (2006). Even conservation rules are made to be broken: implications for biodiversity. *Environmental Management, 37*(2), 162-169.

Rotherham, I. D. (2007). The implications of perceptions and cultural knowledge loss for the management of wooded landscapes: a UK case-study. *Forest Ecology and Management, 249*(1-2), 100-115. https: //doi.org/10.1016/j.foreco.2007.05.030

Rotherham, I. D. (2014). *Eco History: A short history of conservation and biodiversity*. Cambridge, UK: The White Horse Press.

Rotherham, I. D. (2015). Bio cultural heritage and biodiversity – Emerging paradigms in conservation and planning. *Biodiversity and Conservation, 24,* 3405–3429. https: //doi.org/10.1007/s10531-015-1006-5

Rozzi, R., Massardo, F., Anderson, C., Heidinger, K., Silander, J. A.Jr. (2006). Ten principles for biocultural conservation at the southern tip of theAmericas: The approach of the Omora Ethnobotanical Park. *Ecologyand Society, 11*(1), 43. https: //doi.org/10.5751/ES-01709-110143

Samson, C. 2003. A *way of life that does not exist.* Canada: ISER.

Samson, C. and J. Pretty. (2006). Environmental and health benefits of hunting diets and lifestyles for the Innu of Labrador. *Food Policy,* 31(6), 528-553. https: // doi.org/10.1016/j.foodpol.2006.02.001

Sangha, K. K., and Russell-Smith, J. (2017). Towards an indigenous ecosystem services valuation framework: a north Australian example. *Conservation and Society, 15*(3), 255-269.

Sangha, K. K., Preece, L., Villarreal-Rosas, J., Kegamba, J. J., Paudyal, K., Warmenhoven, T., and RamaKrishnan, P. S. (2018). An ecosystem services framework to evaluate Indigenous and local peoples' connections with nature. *Ecosystem services, 31,* 111-125. https: //doi.org/10.1016/j.ecoser.2018.03.017

Schaaf, T., Lee C. *editors. (*2006*)* Conserving cultural and biological diversity: the role of sacred natural sites and cultural landscapes. *Proceedings of the Tokyo Symposium. UNESCO, Paris.*

Singh, B.P. (2017). Biodiversity, tribal knowledge and life in India. *Environment and Social Psychology,* 2(1), 1–10. http: //dx.doi.org/10.18063/ESP.2017.01.001.

Singh, R. K., Hussain, S. M., Riba, T., Singh, A., Padung, E., Rallen, O., and Bhardwaj, A. K. (2018). Classification and management of community forests in Indian Eastern Himalayas: implications on ecosystem services, conservation and livelihoods. *Ecological Processes, 7*(1), 1-15. https: //doi.org/10.1186/s13717-018-0137-5

Skutnabb-Kangas, T., Maffi, L., and Harmon, D. (2003). *Sharing a world of difference: the earth's linguistic, cultural and biological diversity*. Unesco.

Smith DJ, Pilgrim S, Cullen L, (2007) Coral Reefs and People. Chapter 35. In Pretty, J., Ball, A., Benton, T., Guivant, J., Lee, D., Orr, D., Pfeffer, M., Ward, H. (Eds.), SAGE Handbook on Environment and Society. 500-516. SAGE, London

Smith, E. A. (2001). On the coevolution of cultural, linguistic, and biological diversity. *On biocultural diversity: Linking language, knowledge, and the environment,* 95-117.

Stolton, S., Dudley, N., and Randall, J. (2008). *Natural Security: protected areas and hazard mitigation, arguments for protection.* World Wide Fund for Nature (WWF). http: //lib.riskreductionafrica.org/bitstream/handle/123456789/497/ Natural per cent 20security. per cent 20protected per cent 20areas per cent 20and per cent 20hazard per cent 20mitigation.pdf?sequence=1

Sukhdev, P., Wittmer, H., and Miller, D. (2014). The Economics of Ecosystems and Biodiversity (TEEB): Challenges and Responses. In D. Helm, and C. Hepburn (Eds.), *Nature in the Balance: The Economics of Biodiversity*. Oxford, uk: Oxford University Press.

Sutherland, W. J. (2003). Parallel extinction risk and global distribution of languages and species. *Nature, 423*(6937), 276-279. https: //doi.org/10.1038/nature01607

TEEB, (2010). In: Kumar, Pushpam (Ed.), The Economics of Ecosystems and Biodiversity: Ecological and Economic Foundations. Earthscan, London and Washington, p. 456

The World Bank Operational Manual (1991) Operational Directive. http: //www.ifc.org/wps/wcm/connect/835cc50048855270ab94fb6a6515bb18/OD420_IndigenousPeoples.pdf?MOD=AJPERES

UNESCO. (2000). Solving the puzzle: the ecosystem approach and biosphere reserves. http: //www.unesd oc.unesco.org/image s/0011/00119 7/11979 0eb.pdf

UNESCO (2009). World Report Investing in Cultural Diversity and Intercultural Dialogue, 7 place de Fontenoy 75352. Paris 07 SP, France.

UNESCO. (2018). What is Intangible Cultural Heritage? Retrieved from https: //ich.unesco.org/en/what-is-intangible-heritage-00003

UNESCO: COMEST. (2015). Ethical Principles for Climate Change: Adaptation and Mitigation: Report of COMEST. Paris. https: //unesdoc.unesco.org/ark: /48223/pf0000234529

United Nations. (1992). the Convention on Biological Diversity (CBD). Rio Conference. https: //www.cbd.int/doc/legal/cbd-en.pdf

World Bank. (2020). *Investing in Nature Pays off for People and Biodiversity*. https: //www.worldbank.org/en/news/immersive-story/2020/05/21/investing-in-nature-pays-off-for-people-and-biodiversity?cid=ECR_FB_worldbank_EN_EXTP_R

World Conference on Indigenous Peoples (2014). INTER-AGENCY SUPPORT GROUP ON INDIGENOUS PEOPLES' ISSUES THEMATIC PAPER. https: //www.un.org/en/ga/69/meetings/indigenous/pdf/IASG per cent 20Thematic per cent 20Paper per cent 20- per cent 20Health per cent 20- per cent 20rev1.pdf

WWF. (2013). Working with Indigenous and Local Knowledge Systems for the Conservation and Sustainable Use of Biodiversity and Ecosystem Services. An Analysis of Selected Case Studies from WWF Projects Worldwide as a Contribution to IPBES-2. WWF Information. http: //awsassets.panda.org/downloads/wwf_ipbes_ilk_information_2013.pdf

Chapter 14

Diversity of Spiders in different Urban Gradients of Kalaburagi City, Karnataka, India

T. Supriya[1], Shashikanth H. Majagi[1] and K. Vijaykumar[2]*

Department of Zoology,
Vijaynagar Sri Krishnadevaraya University, Ballari – 583 105, Karnataka
[2]Department of Zoology,
Gulbarga University, Kalaburagi – 585 106, Karnataka
**Corresponding author e-mail: smajgi@rediffmail.com*

Effects of urbanization on spiders (Araneae) were studied between March 2016 to March 2017along the urban-suburban–exurban-agriculture land-parks/garden gradient in Kalaburagi city, Karnataka. The total species richness was 21 species, 21 genera and 12 families. The family Salticidae made up 14.28 per cent the species followed by Araneidae, Lycosidae, Theridiidae and Thomisidae each made up 9.5 per cent total occurrence. We found the highest diversity of spiders in exurban and agriculture lands followed by suburban and least in Urban and park/garden. The species accumulation curves are still a useful tool to evaluate sampling quality and compare inventories of mega-diverse groups, such as spiders. Some aspects of spider niche specialization are poorly documented in study area, specifically how urban development affects the diversity and abundance of spider. Our study stresses the spider diversity and abundance may be influenced by alterations of habitats by humans through urbanization.

Keywords: *Spiders, Anthropogenic activity, Urban gradients, Kalaburagi urban.*

Introduction

All cities are originally developed by the suburbanized habitats. During course of expansion suburbans substitute urban by matrix of buildings, solid pavement and dense road network with high population density (Gilbert 1989). Patterns of city growth in configuration with variety of effects, such as pollution, drainage, watercourse diversion, fragmentation, habitat loss; alteration in temperature, precipitation regimes, nutrient cycle catalysed by anthropogenic activity. All these upshots of anthropogenic activities are similar in different parts of the world, thus cities tend to be more similar to each other than compared to the environment where they are immersed (McIntyre 2000, Alberti 2005, Pickett *et al.,* 2011). Although urbanization is predominantly influence on plants and animals species diversity, abundance, population dynamicity and Phenologyin both negative and positive directions (Faeth *et al.,* 2011, Aronson *et al.,* 2014, Lowe *et al.,* 2014, Meineke *et al.,* 2017). Some species exhibited favourable responses to urbanization by occur at even higher abundances than in less disturbed systems (Fischer and Knutti 2015) while many native and specialist species reacted negatively and have disappeared from these areas (Eversham *et al.,* 1996, Chudzicka and Skibinska 1998).

Study of ecological patterns and their influence on distribution of different species within urban environment plays momentous role in the field of conventional biology (Miller and Hobbs 2002). Scrutiny of ecological disturbance such as alteration of plant and animal species diversity and abundance from urban through suburban to rural environment provides figure and fact about effect of urbanization on ecological system (McDonnell and Pickett 1990, McDonnell *et al.,* 1993, Niemelä *et al.,* 2000). It is vital to pick out organisms that have essential ecosystem functions, such as the megadiverse arthropods that can indicate drastic biodiversity losses as well as anticipate impacts to the ecosystem processes (Walker *et al.,* 1992, Gerlach *et al.,* 2013).Another reason for studying arthropods in urban habitats is that they are important in agriculture and gardens as pests or biological control agents and in medicine (McIntyre 2000).

Among arthropods one suitable target group is spiders (Araneae), commonly associated with urban habitats and human dwellings due to their high dispersal ability. Therefore, the differences in diversity and abundance of spiders may reflect the changes in trophic structure in human-altered systems (Shochat *et al.,* 2008).This large group of spiders may respond differently to land use, intensity and type of habitat management practices and sometime alternatively taking advantage of built structures in the absence of plants (Dahirel *et al.,* 2017),hence they may be considered as bio-indicators of habitat disturbance and are extremely sensitive to environmental changes that act on habitat structure, causing significant modifications on the distribution of the species (Wise 1993, Hendrickx *et al.,* 1998, Skerl and Gillespie 1999, Cardoso *et al.,* 2004).

Domestic gardens and abandoned plots may play a vital role in maintaining the diversity of invertebrates in cities (Smith *et al.*, 2006).Human populations continue to modify natural habitats, it is important to understand how anthropogenic disturbance affects the population dynamics of native species in order to mitigate significant changes to the function of urban ecosystems (Lowe *et al.*, 2014).Hence in this study we sought to assess diversity of spiders in different degree of urbanization in the city of Kalaburagi : urban, suburban, exurban, agriculture fields and parks/ gardens. Our aim was to articulate,how spider assemblages are influenced by an ongoing process of urbanisation. Qualitative and quantitative inventory carried out in this study provides knowledge and checklist of spiders in the city and this effort influence conservation plans for this anthropic ecosystem.

Materials and Methods

Study Area

The study was carried out on different urban zones of Gulbarga city (17°32′N Latitudes and 76°83′ E Longitudes) of Karnataka, India. It is situated at elevation 474 meters above sea level. This entire district is mainly composed of Deccan trap. The vast expanse of treeless plateau is interspersed with river plain and watersheds. Kalaburagi city is located on an undulating plain, presenting a vast stretch of black cotton soil. Being a regional headquarters of government, it is an important city in the Northern part of Karnataka situated 623 km away from Bangalore and 220 km from west of Hyderabad, Kalaburagi is well connected by rail and road.

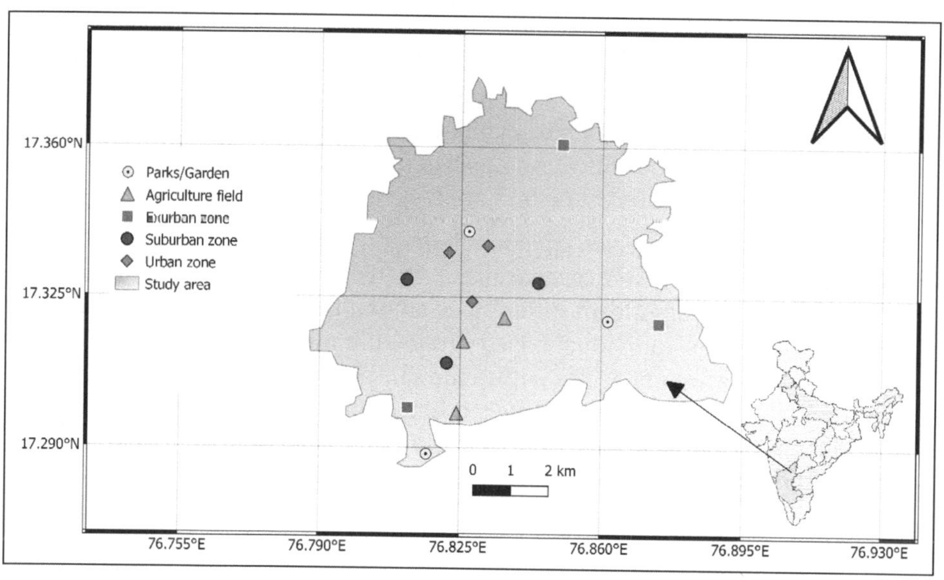

Figure 14.1: Location of Sampled Sites in each Urban Gradients of City Kalaburagi, Karnataka, India.

Sampling Site

This survey covers three different urban zones of Kalaburagi city includes Urban zone: situated at centre of the city consisted of matrix of buildings receiving heavy disturbances from transportations, automobile emissions and human crowed; Suburban: Low density urbanization; Exurban: borders of city where scares of houses, no urbanization; Agriculture fields: vegetable fields within suburban and exurban zones; Parks/Gardens: artificially constructed habitats comprises flora mainly of ornament trees.

Sampling Methods

For this study, we categorized 4 sampling zones: urban, suburban, exurban, agriculture fields within city and park/gardens within city. The distance between the edges of these sites was at least 500 m so that they could be considered independent at the site level. Sampling was conducted during March 2016 – March 2017. These specimens were collected from insides and outsides the homes, terries, stones near homes, under the flowering pots, gardens, walls, courtyard, stairwells, windows, adjacent vegetations manually using forceps and placed vials against movement of spiders allow that to enter in the vial, then preserved in of 70 per cent ethanol. Spiders were sorted and identified in the laboratory under under Carl-Zeiss Stemi 2000-c Stereo-Zoom microscope mounted with Axio Cam ERc5s camera and Olympus SZ61 Stereo-Zoom microscope mounted with Slides SLI500 camera. All collected adult spiders were identified to species level using original published descriptions and revisions for each family group and nomenclature follows the World spider catalogue (Platnick, 2012) and consulting taxonomic specialist. The female epigyne and the males palpi of adult specimens were dissected, and 10 per cent KOH was used to remove the soft internal tissue of the epigyne. The immature was excluded from the analysis to avoid over- or underestimation of richness patterns, because their identification is extremely difficult (Sørensen 2004) or not possible since reproductive characters are considered to separate some species.

Data analysis: To measure specific biodiversity of all species in the sample, the different biodiversity indices was used. ANOVA tests were used to determine whether there were significant differences among the abundances of species in different urban zones. Individual based rarefaction models explicitly account for the relative abundance of species within the sample pool. To assess whether local (site) abundances and richness differed between urbanization levels (habitat types), richness, C and CV were also estimated for each site. All the data analysis done with the soft-ware Paleontological Statistics (PAST) v4.05.

Results

The species richness in associated with different urban zone of Kalaburagi city, Karnataka was 21 species, 21 genera and 12 families between March 2016 to March 2017 in different urban zones (Table 14.1). Concerning the quality and total sampling effort in different urban gradients, the coefficient of determination was R^2

=0.957, has strong correlation since value is close to 1. The family Salticidae made up 14.28 per cent the species followed by Araneidae, Lycosidae, Theridiidae and Thomisidae each made up 9.5 per cent total occurrence. exurban and agriculture field (17 species) held a significantly higher number of species compared to other urban zones, mediate in suburban (15 species) and least in urban and Park/garden. *Plexippus paykulli, Stenaelurillus arambagensis* are present in all urban zones.

Table 14.1: The Checklist of Spiders Reported during this Study from different Urban Gradients of Kalaburagi City, Karnataka

Sl. No.	*Family*	*Genus*	*Species*	*UR*	*SU*	*EU*	*AF*	*PG*
1	Araneidae Clerck, 1757	Araneus	*Araneus inustus* Koch, 1871	✗	✗	✓	✓	✓
		Argiope	*Argiope anasuja* Thorell, 1887	✗	✗	✗	✓	✗
2	Corinnidae Karsch, 1880	Castianeira	*Castianeira zetes* Simon, 1897	✓	✓	✓	✓	✗
3	Dictynidae Pickard-Cambridge, 1871	Brigittea	*Brigittea civica* (Lucas, 1850)	✗	✓	✗	✗	✗
4	Filistatidae Ausserer, 1867	Filistata	*Filistata napadensis* Patel, 1975	✗	✓	✓	✗	✗
5	Gnaphosidae Pocock, 1898	Gnaphosa	*Gnaphosa kailana* Tikader, 1966.	✓	✓	✗	✓	✓
		Zelotes	*Zelotes tenuis* (Koch, 1866)	✗	✗	✓	✓	✓
6	Lycosidae Sundevall, 1833		*Pardosa birmanica* Simon, 1884		✓	✗	✗	
			Arctosa himalayensis Tikader and Malhotra, 1980	✗		✓	✓	
7	Oxyopidae Thorell, 1870		*Oxyopes sertatoides* Xie and Kim, 1996	✗	✗	✓	✓	✓
			Peucetia yogeshi Gajbe, 1999		✗	✓	✓	
8	Philodromidae Thorell, 1870		*Philodromus* sp. 1 Walckenaer, 1826	✓	✓	✓	✓	
9	Pholcidae Koch, 1850		*Crossopriza lyoni* (Blackwall, 1867)	✗	✓	✓	✓	
10	Salticidae Blackwall, 1841		*Aelurillus* sp. 1 Simon, 1884		✓	✓	✓	
			Plexippus paykulli (Audouin, 1826)	✓	✓	✓	✓	
			Stenaelurillus arambagensis (Biswas and Biswas, 1992)	✓	✓	✓	✓	✓
			Cyrba ocellata (Kroneberg, 1875)	✓	✓	✓	✓	

Sl. No.	Family	Genus	Species	UR	SU	EU	AF	PG
11	Theridiidae Sundevall, 1833		*Steatoda* sp. 1 Sundevall, 1833		×	✓	✓	
			Theridion indicum Tikader, 1977	×	✓	✓	✓	
12	Thomisidae Sundevall, 1833		*Thomisus* sp. Walckenaer, 1805		✓	×	×	
			Ozyptila sp. Simon, 1864		✓	✓	✓	

UR: Urban; SU: Suburban; EU: Exurban; AF: Agriculture field; PG: Park/Garden.

We used different diversity indices to calculate diversity (Table 14.2). All diversity indices indicated that- diversity was high at agriculture fields and exurban zones; species abundance was highest in agriculture lands and suburban zones; Shannon Index is higher in Agriculture field and less in urban zone; Species are more evenly distributed in agriculture field and park/Garden. Simpson diversity index shows highest value in exurban and Park indicates more diversity of species. All urban habitats had C values close to 100 per cent (83 per cent, 92 per cent, 94 per cent, 96 per cent and 88 per cent for urban, suburban, exurban, agriculture field and Park/garden habitats respectively) indicating a good representation of the species in the samples. There were Fisher's alpha (ANOVA F=33.56, df=4, P=0.004), simpson index (ANOVA F=25.63, df=4.07, P=0.006). Simpson index and Fisher's alpha were moderately correlated across all urban zones (Pearson correlation, $r^2 = 0.33$, $P < 0.05$).

Table 14.2: Total Specific Richness (S) of Spiders and Diversity Index for Each Sampling Zone in the Kalaburagi City, Karnataka, India

	Urban	Suburban	Exurban	Agriculture	Park/Garden
Taxa	11	15	17	17	11
Individuals	41	70	53	67	23
Dominance D	0.125	0.077	0.055	0.06	0.05
Simpson1-D	0.87	0.92	0.94	0.93	0.94
Shannon_H	1.98	2.42	2.55	2.57	2.14
Evenness e^H/S	0.66	0.75	0.75	0.77	0.77
Chao-1	11.75	15	23	17	11

Box plot indicates that there is a difference among the urban zone (Figure 14.2). Among different urbanization gradients suburban and agriculture land shows more dispersal of individuals of spider species but individuals in Park/Garden are less dispersed. Urban is left skewed, agriculture and exurban are right skewed and suburban and Parks/Garden are shows normal distribution. Rarefraction curve for different habitats of urban gradients appears similar between exurban and agriculture; suburban and park/garden, but suburban and park/garden level off earlier whereas agriculture and suburban continue to increase moderately indicates species acquisition with sampling effort (Figure 14.3).

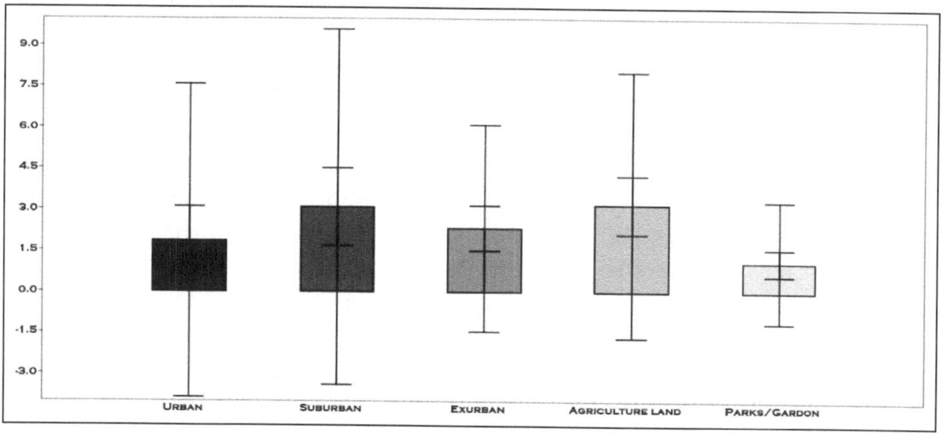

Figure 14.2: Box Graph Showing Dispersal of Spiders Species among different Urban Gradients.

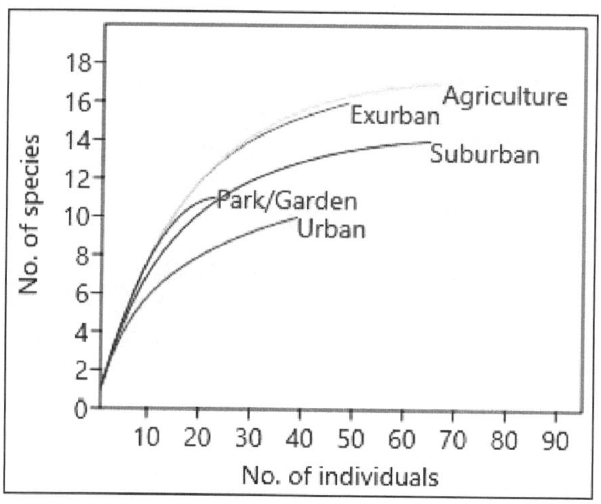

Figure 14.3: Species Refraction Curve for different Urban Gradients of Kalaburagi City, Karnataka.

Discussion

This preliminary survey of spiders along different urban habitat gradients has yielded an impressive result of 21 taxa of spiders (21 genera and 12 families) from a very limited study area and with a relatively low sampling effort. The highest diversity of spiders was reported in exurban and agriculture fields followed by suburban and least in Urban and park/garden. These variation may be due to the alteration of microclimates within each zone. The abundances and species richness of the spider taxa were significantly different across the various urbanisation gradients (urban, suburban, exurban agriculture lands and parks/garden in this study) shows different tolerance and dispersal ability among different urban zones. Abundance of spider is less in urban zones as it consists of matrix building and receives more anthropogenic activity alters temperature and humidity which occur in the urban matrix. Our study did not reveal physical parameters effects of urbanization on spider distribution. Even though ballooning in spiders allow them to disperse easily, thin cuticle of spiders make them more sensitive to thermal and humidity conditions (Lövei *et al.,* 2019). Green space in urban zones plays very important role in preservation of biodiversity. However, most maintenance and management operations harm biodiversity in urban green spaces thereby facilitating the maintenance of natural processes of decomposition and nutrient cycling (Lövei *et al.,* 2019).Obedient to this, in our study we encountered more spider species in agriculture lands compared to man-made artificial habitats park/ garden within city indicates natural habitats are more suitable and preferable by spiders.

Many studies shows that the richness of spider species is negatively affected by urbanization in general, mainly by groups of web weavers, as they are more sensitive to changes in habitat structure (vegetation). Contrary to this, our study reveals that urban zone of Kalaburagi city has less species diversity compared to other habitats. Our study holds up the statement that structure of urban landscapes has a significant influence in species richness, functional richness, and community composition (Melo *et al.,* 2021).

Other studies, however, have found that urbanization has significant positive as well as negative effects on various taxonomic groups (Argañaraz *et al.,* 2017, Lövei *et al.,* 2019, Melo *et al.,* 2021). Some studies revels that urbanization has no harmful consequence on the spiders (Magura *et al.,* 2010, Melliger *et al.,* 2018). Depending on number or methods of sampling efforts, the species richness can be different in the same area (Dias *et al.,* 2006).As we have done low sampling effort, we suggest that further studies should use several sampling methods and further research is needed in concerning to effect of urbanization and human disturbance intensity on structural diversity, stand size, age in critical urban gradients. From this study we recommend that it is necessary to urban maintenance policy makers to consider effect of small-scale environmental factors, litter layers and ground vegetation on different taxonomic and functional groups while implementing urban planning in protection of maximize conservation value of urban green areas (Melliger *et al.,* 2018).

Acknowledgment

Authors are thankful to Gulbarga University, Kalaburagi for financial assistance. Our special thanks extended to Principal Dr. Subhash Badange J. D. Patil Sangludkar Mahavidyalaya Daryapur, Maharashtra for providing lab facility for identification.

We wish to acknowledge. We are grateful to Dr. Shankrappa Hatti, Dr. Anita S Majagi, Dr. Heena Mubeen, Premanjali, for their important contributions, comments and suggestions to improve this work.

REFERENCES

Alberti, M. 2005. The effects of urban patterns on ecosystem function. International regional science review 28: 168-192.

Argañaraz, J. P., V. Radeloff, A. Bar-Massada, G. I. Gavier-Pizarro, C. M. Scavuzzo, and L. M. Bellis. 2017. Assessing wildfire exposure in the wildland-urban interface area of the mountains of central Argentina. Journal of environmental management 196: 499-510.

Aronson, M. F., F. A. La Sorte, C. H. Nilon, M. Katti, M. A. Goddard, C. A. Lepczyk, P. S. Warren, N. S. Williams, S. Cilliers, and B. Clarkson. 2014. A global analysis of the impacts of urbanization on bird and plant diversity reveals key anthropogenic drivers. Proceedings of the royal society B: biological sciences 281: 20133330.

Cardoso, P., I. Silva, N. G. de Oliveira, and A. R. Serrano. 2004. Indicator taxa of spider (Araneae) diversity and their efficiency in conservation. Biological Conservation 120: 517-524.

Chudzicka, E., and E. Skibinska. 1998. Diversity of reactions of insect communities as a response to anthropogenic pressure. MEMORABILIA ZOOLOGICA-INSTYTUT ZOOLOGICZNY POLSKA AKADEMIA NAUK: 13-30.

Dahirel, M., J. Dierick, M. De Cock, and D. Bonte. 2017. Intraspecific variation shapes community level behavioral responses to urbanization in spiders. Ecology 98: 2379-2390.

Dias, S. C., A. D. Brescovit, E. C. Couto, and C. F. Martins. 2006. Species richness and seasonality of spiders (Arachnida, Araneae) in an urban Atlantic Forest fragment in Northeastern Brazil. Urban Ecosystems 9: 323-335.

Eversham, B. C., D. B. Roy, and M. G. Telfer. 1996. Urban, industrial and other manmade sites as analogues of natural habitats for Carabidae. Pages 149-156 *in* Annales Zoologici Fennici. JSTOR.

Faeth, S. H., C. Bang, and S. Saari. 2011. Urban biodiversity: patterns and mechanisms. Annals of the New York Academy of Sciences 1223: 69-81.

Fischer, E. M., and R. Knutti. 2015. Anthropogenic contribution to global occurrence of heavy-precipitation and high-temperature extremes. Nature Climate Change 5: 560-564.

Gerlach, J., M. Samways, and J. Pryke. 2013. Terrestrial invertebrates as bioindicators: an overview of available taxonomic groups. Journal of Insect Conservation 17: 831-850.

Gilbert, O. 1989. The ecology of urban habitats–Chapman and Hall. London/New York: 1-369.

Hendrickx, F., J.-P. Maelfait, W. Muylaert, and M. Hoffmann. 1998. Spider distribution patterns along the tidal River Scheldt (Belgium). Pages 285-291 *in* Proceedings of the 17th European Colloquium of Arachnology. Dorset Press, Dorchester, GB.

Lövei, G. L., R. Horváth, Z. Elek, and T. Magura. 2019. Diversity and assemblage filtering in ground-dwelling spiders (Araneae) along an urbanisation gradient in Denmark. Urban Ecosystems 22: 345-353.

Lowe, E. C., S. M. Wilder, and D. F. Hochuli. 2014. Urbanisation at multiple scales is associated with larger size and higher fecundity of an orb-weaving spider. PLoS One 9: e105480.

Magura, T., R. Horváth, and B. Tóthmérész. 2010. Effects of urbanization on ground-dwelling spiders in forest patches, in Hungary. Landscape Ecology 25: 621-629.

McDonnell, M. J., and S. T. Pickett. 1990. Ecosystem structure and function along urban rural gradients: an unexploited opportunity for ecology. Ecology 71: 1232-1237.

McDonnell, M. J., S. T. Pickett, and R. V. Pouyat. 1993. The application of the ecological gradient paradigm to the study of urban effects. Pages 175-189 Humans as components of ecosystems. Springer.

McIntyre, N. E. 2000. Ecology of urban arthropods: a review and a call to action. Annals of the Entomological Society of America 93: 825-835.

Meineke, E. K., A. J. Holmquist, G. M. Wimp, and S. D. Frank. 2017. Changes in spider community composition are associated with urban temperature, not herbivore abundance. Journal of Urban Ecology 3: juw010.

Melliger, R. L., B. Braschler, H.-P. Rusterholz, and B. Baur. 2018. Diverse effects of degree of urbanisation and forest size on species richness and functional diversity of plants, and ground surface-active ants and spiders. PLoS One 13: e0199245.

Melo, T., E. Moreira, M. Lopes, A. Andrade, A. Brescovit, M. Peres, and J. Delabie. 2021. Influence of urban landscape on ants and spiders richness and composition in forests. Neotropical Entomology 50: 32-45.

Miller, J. R., and R. J. Hobbs. 2002. Conservation where people live and work. Conservation biology 16: 330-337.

Niemelä, J., J. Kotze, A. Ashworth, P. Brandmayr, K. Desender, T. New, L. Penev, M. Samways, and J. Spence. 2000. The search for common anthropogenic impacts on biodiversity: a global network. Journal of Insect Conservation 4: 3-9.

Pickett, S. T., M. L. Cadenasso, J. M. Grove, C. G. Boone, P. M. Groffman, E. Irwin, S. S. Kaushal, V. Marshall, B. P. McGrath, and C. H. Nilon. 2011. Urban ecological systems: Scientific foundations and a decade of progress. Journal of environmental management 92: 331-362.

Platnick, N. I. 2012. The world spider catalog. Version 12.5. The American Museum of Natural History.

Shochat, E., W. Stefanov, M. Whitehouse, and S. H. Faeth. 2008. Urbanization and spider diversity: influences of human modification of habitat structure and productivity. Pages 455-472 Urban Ecology. Springer.

Skerl, K. L., and R. G. Gillespie. 1999. Issues in Spider Conservation. Kluwer.

Smith, R. M., P. H. Warren, K. Thompson, and K. J. Gaston. 2006. Urban domestic gardens (VI): environmental correlates of invertebrate species richness. Biodiversity and Conservation 15: 2415-2438.

Walker, H. M., L. K. Irvin, J. Noell, and G. H. Singer. 1992. A construct score approach to the assessment of social competence: Rationale, technological considerations, and anticipated outcomes. Behavior Modification 16: 448-474.

Wise, D. 1993. Spiders in ecological webs Cambridge University Press. New York ISBN-13: 978-0521.

Chapter 15

An Efficient *In vitro* Approach for Callogenesis Mediated from the Plants *Tabernaemontana heyneana* Wall.

D.J. Manasa[1,2], K.R. Chandrashekar[1,3], N. Bhagya[3], D.J. Madhu Kumar[4], M.D. Reshma[5] and M. Govindappa[2]*

*[1]Department of Applied Botany,
Mangalore University, Mangalagangothri, Mangalore – 574 199, Karnataka
[2]Department of Studies in Botany, [5]Department of Studies in Zoology,
Davangere University, Shivagangothri, Davangere – 577 007, Karnataka
[3]Yenepoya Research Centre,
Yenepoya (Deemed to be University), Mangalore – 575 018, Karnataka
[4]Faculty of Pharmacy,
Ramaiah Institute of Applied Sciences, Bangalore, Karnataka
Corresponding author e-mail: manasadj310@gmail.com

The study was aimed to develop an efficient, rapid and large-scale in vitro callus induction of Tabernaemontana heyneana Wall. which is an highly medicinal endemic plant of Western Ghats. The study helps in the conservation and restoration of this plant which relieves pressure on the wild plants for the extraction and isolation of medicinally important bioactive principles used in the treatment of various ailments. An efficient surface sterilization protocol was developed for Tabernaemontana heyneana for the aseptic in vitro culturing. A highest percentage (76.33 per cent) of aseptic cultures were obtained by the treatment of 2 per cent bavistin (1 h 30 min) followed by 1 per cent $HgCl_2$ (5min) + 70 per cent ethanol (1min) for T. heyneana explants. About 98 per cent of callus induction frequencies were obtained in the MS-medium supplemented with 3mg/l 2, 4-D + 2mg/l Kn which is comparatively higher than *any other combinations in T. heyneana. This protocol allows the mass production of callus of Tabernaemontana heyneana for commercial uses, and for conservation purposes; calluses can be used to establish suspension cultures in prospecting for bioactive compounds.*

***Keywords**: Tabernaemontana heyneana Wall., Callus induction, In vitro, 2,4-D, Kn.*

Introduction

All over the globe, medicinal plants are mainly used as crude drugs. Number of potent and active substances from plants are employed as isolated compounds, including many alkaloids such as morphine (pain killer), codeine (antitussive), ephedrine (stimulant), ajmaline (anti-rhythmic), quinine (anti-malarial), reserpine (anti-hypertensive), galanthamine (acetylcholine esterase inhibitor), scopolamine (travel sickness), berberine (psoriasis), caffeine (stimulant), capsaicin (rheumatic pains), colchicine (gout), pilocarpine (glaucoma) and various types of cardiac glycosides (heart insufficiency) (Wink *et al.,* 2005).

The capacity for plant cell, tissue and organ cultures to produce and accumulate many of the similar significant chemical compounds as the parent plant in nature has been realized almost since the initiation of *in vitro* technology. The robust and flourishing demand in today's marketplace for natural, renewable products has focused the attention on *in vitro* plant materials as potential factories for secondary bioactive principles, and has paved the way for new research on these lines. However, it is not only commercial significance that drives the research initiatives but, plant tissue culture techniques play a crucial role in the search of alternatives for the formulation of desirable medicinal compounds from plants (Naik and Al-Khayri2017). The deliberate stimulation of defined chemical products within carefully regulated *in vitro* cultures provides an excellent forum for in-depth investigation of biochemical and metabolic pathways, under highly controlled micro environmental regimes.

There are various specific advantages for producing a beneficial secondary product *in vitro*, rather than *in vivo*. These include: a) Production can be highly principled, simpler, more certain and independent of seasonal variations b) Metabolite production can be efficiently accelerated when compared to isolation from an organised complex tissues of plant c) Compounds produced *in vitro* can be directly parallel to the compounds in the plant d) Interrupting compounds that occur in the wild plant can be avoided in *in vitro* culture e) Tissue and cell cultures can yield a source of defined standard phytochemicals in large volumes f) Tissue and cell cultures are a potential model to test elicitation g) Cell cultures can be radio labelled, such that the accumulated secondary products, when provided as feed to laboratory animals, can be traced metabolically. Obnoxious odours or flavors associated with the crop plant can be altered or eliminated *in vitro*. Plant cell culture eliminates potential political boundaries or geographic barriers to the production of a crop, such as the restriction of natural rubber production to the tropics or anthocyanin pigment production to climates with high light intensity. When a valuable product is found in a wild or scarce plant species, intensive cell culture is a practical alternative to wild collection of plant materials.

Callus is a mass of unorganised and undifferentiated parenchyma cells which could be initiated in a medium, conducive to undifferentiated cell proliferation supplemented with plant growth regulators like auxins (2,4-D, NAA, IAA, IBA), cytokinins (Kinetin, 6-benzylaminopurine, zeatin, 1,3-diphenylurea, thiadiazuron), gibberlins *etc.* Although, callogenesis and isolation of secondary metabolites is well documented in the members of Apocynaceae, but *T. heyneana* have not been studied.

Figure 15.1: ***Tabernaemontana heyneana*** **Wall.**
(A) Habit-Small deciduous tree; (B) Inflorescence-Terminal corymbose; (C) Flower-White fragrant; (D) Fruit-Yellow orange coloured follicle.

Tabernaemontana heyneana Wall. (Syn. *Tabernaemontana alternifolia*) known as Maddarassagida/Madlemara/Nagarkuda in Kannada usually grown in the open forests of the Western Ghats from Konkan (Maharashtra) to southwards through Kerala up to 900 meter elevation. A small deciduous tree that grows upto

9m tall with rough grey bark (Figure 15.1A). Leaves up to 18 × 6 cm, elliptic to lanceolate, acuminate at apex, base acute, petiole 1cm long. Inflorescence terminal corymbosepeduncled cymes (Figure 15.1B), pedicels 0.6 - 2 cm long. Calyx - lobes broadly ovate, apiculate. Corolla white, tube 1.5 - 3 cm long, lobes broad falcate, with crisped margins (Figure 15.1C). Follicles 2.5 - 3.5 cm long, bright orange -yellow, beaked, usually with 2 sharp sides (Figure 15.1D). Plant is endemic to Western Ghats which flowers in the month of March and April.

The plant is known to possess antimicrobial activity against skin and venereal diseases; also effective in the treatment of respiratory problems, nervous disorders and various other diseases (Ignacimuthu and Ayyanar 2005; Ignacimuthu *et al.*, 2006). Different types of alkaloids *viz.* tabernoxidine, coronaridine and vocangine were reported from *T. heyneana* (Roy *et al.*, 2002). Compounds like rutin, quercetin and few phenolic acids have also been identified from the leaves of *T. heyneana* (Sathishkumar *et al.*, 2008).

The plant is known for the presence of various secondary metabolites with therapeutic potential and are being used in the traditional medicinal system hence, the demand for these plants have drastically increased. There are no reports on the *in vitro* approach on *T. heyneana,* therefore an attempt has been made on the production of undifferentiated *in vitro* cell cultures (callus cultures).

Methodology

Standardization of Surface Sterilization

The leaves and shoot segments were excised from young and healthy plants of *T. heyneana* [MU/AB/DJM-01] collected from natural forests of Dakshina Kannada district (12.8158 °N and 74.9241 °E), Karnataka, India. Taxonomic identification was done following the Flora of Madras Presidency (Gamble 1958). Voucher specimens have been deposited in the herbarium of Dept of Applied Botany, Mangalore University.

The mature, young and youngest leaves, petiole along with shoot apical meristems and nodal stem cuttings were collected to find out the suitable explants for *in vitro* culture. The collected leaves and shoot segments were washed thoroughly under running tap water for 60 min to remove the surface micro organisms and dust particles. The explants were treated with 2 per cent bavistin (1 h and 30 min) for *T. heyneana* and 1 per cent bavistin (1h) followed by rinsing in distilled water for 3-4 times. The explants were then surface disinfected with a powerful sterilizing agent 70 per cent alcohol (30s, 1 and 2min) followed by immersion in different surface sterilizing agents like mercuric chloride ($HgCl_2$w/v) and sodium hypochlorite (NaOCl v/v) in a laminar air flow cabinet. The leaf and stem explants were exposed to NaOCl (1, 1.4 per cent) and $HgCl_2$ (0.1, 1, 2 per cent) for different durations like 5, 15, 30 min and 1, 2, 3, 4, 5, 10 min respectively. After each step, the explants were washed with sterile distilled water. All the above mentioned sterilizing agents were used alone or in combination to get the optimum results of surface sterilization of the explants. Later, the explants were cut into 1 cm^2 and inoculated on to the basal Murashige and Skoog medium (1962). The cultures were maintained at 16/8 h photoperiod under an illumination of 3000 -3200 lux light intensity. After 30 days, the percentage of explants survival, browning, fungal and bacterial contamination was recorded. After 30 days of observation the explants remaining fresh with specific treatment of sterilizing agents were selected for the induction of callus.

Preparation of Media and Culture Conditions

The composition and preparation of stock solutions for MS - medium is given in Table 15.1. Appropriate quantities of various stock solutions and plant growth regulators were pipetted out and stirred with 400 ml distilled water. After adding 3 per cent of sucrose, the pH was adjusted to 5.7 with 0.1 N NaOH/0.1 N HCl. The agar (0.7 per cent) was added and the final volume was made upto 1l with distilled water. The medium was sterilized in an autoclave at 15 psi for 15 min and cooled.

Table 15.1: MS Media Composition

Stock Solution Desig- nation	*Ingredient*	*Weight of Ingredient (mg)*	*Volume of Water Used (ml)*	*Volume of Stock Solution taken for Making 1l of Medium (ml)*	*Final Conc. of the Ingredient in the Medium (mgl^{-1})*
I	**Macronutrient**				
A	NH_4NO_3	16500	500	50	1650
	KNO_3	19000	500	50	1900
B	$MgSO_4 \cdot 7H_2O$	3700	500	50	370
	KH_2PO_4	1700	500	50	170
C	$CaCl_2 \cdot 2H_2O$	4400	500	50	440
II	**Micronutrient**				
	H_3BO_3	620	500	50	6.2
	KI	83	500	50	0.83
D	$Na_2MoO_4 \cdot 2H_2O$	25	500	50	0.25
	$CoCl_2 \cdot 6H_2O$	2.5	500	50	0.025
	$CuSO_4$	2.5	500	50	0.025
E	$ZnSO_4$	860	500	50	8.6
	$MnSO_4$	2230	500	50	22.3
	$Na_2EDTA2H_2O$	373	200	20	37.3
	$FeSO_4 \cdot 7H_2O$	278	200	20	27.8
	Glycine	40	200	20	2
	Nicotinic acid	10	200	20	0.5

Stock Solution Desig-nation	*Ingredient*	*Weight of Ingredient (mg)*	*Volume of Water Used (ml)*	*Volume of Stock Solution taken for Making 1l of Medium (ml)*	*Final Conc. of the Ingredient in the Medium (mgl^{-1})*
F	Thiamine HCl	2	200	20	0.1
	Pyridoxine HCl	10	200	20	0.5
	Myo-inositol	2000	200	20	100
	Sucrose	30000	200	20	30000
	Agar		200	20	8000

Initiation, Growth and Maintenance of Callus

To study the initiation and growth of callus, MS media with varying concentrations of plant growth regulators such as 2,4-D (2,4-dichlorophenoxyacetic acid), NAA (1-Naphthaleneacetic acid), BAP (6 - Benzylaminopurine), Kn (Kinetin) either individually or in combination were used to find out the suitable medium for callogenesis. The leaf and stem explants were used as inocula. Observations like development of callus from mature or immature regions of the explants, nature, texture and colour of the callus were made within three weeks of inoculation. Each treatment combination was assigned to 20 explants with one explant per test tube in 3 replicates. The fully grown calli was subcultured to the MS media supplemented with effective growth regulator combination for the mass production of callus. Once in every 30 to 40 days, the calli were transferred to fresh medium of the same composition. Callus was harvested when the callus cultures turned brown and there was no further development of callus.

Microscopic Observation of Callus

The calli were observed using light microscope by squashing the callus on a clean glass slide with a drop of water/aniline blue stain for their size, shape and cytodifferentiation.

Results

Standardization of Surface Sterilization

The highest percentage of a septic cultures (76.33 per cent) was obtained by treating the *T. heyneana* (Table 15.2) leaf explants with 1 per cent mercuric chloride for 5 min + 70 per cent ethyl alcohol for 1 min. These treatments resulted in a highest percentage of explant survival.

Effect of different sterilants on apical meristems and nodal stem cuttings of *T. heyneana* is given in Table 15.3. The rate of explants survival was very low (10 per cent) that it was impossible to revive the explants even after incubating for 45 days. The harsh treatment with powerful sterilizing agents like $HgCl_2$ and NaOCl

Table 15.2: Influence of different Sterilants on per cent of Browning, Fungal Contamination, Bacterial Contamination and Survival of Leaf Explants in *T. heyneana*

Sterilant Used	*Time of Exposure*	*Browning (Per cent)*	*Fungal Contamination (Per cent)*	*Bacterial Contamination (per cent)*	*Explant Survival (Per cent)*
Sodium hypochlorite (1 per cent)	5min	66.65	77.65	18.43	20.12
Sodium hypochlorite (1 per cent)	15min	76.43	53.42	24.56	38.45
Sodium hypochlorite (1 per cent)	30min	85.54	55.77	15.77	45.62
Sodium hypochlorite (1.4 per cent)	5min	89.32	52.31	23.63	46.52
Sodium hypochlorite (1.4 per cent)	15min	91.25	48.32	36.52	32.65
Sodium hypochlorite (1.4 per cent)	30min	92.35	32.32	51.32	23.65
Mercuric chloride (0.1 per cent) + ethyl alcohol (70 per cent)	2 min + 30 s	16.56	49.65	51.76	49.76
Mercuric chloride (0.5 per cent) + ethyl alcohol (70 per cent)	5 min + 30 s	25.86	16.76	40.45	24.76
Mercuric chloride (1 per cent) + ethyl alcohol (70 per cent)	1 min + 30 s	33.56	23.67	18.54	49.34
Mercuric chloride (1 per cent) + ethyl alcohol (70 per cent)	3 min + 30s	32.67	37.76	25.54	50.65
Mercuric chloride (1 per cent) + ethyl alcohol (70 per cent)	4 min + 30s	46.54	28.65	37.65	60.56
Mercuric chloride (1 per cent) + ethyl alcohol (70 per cent)	5 min + 1 min	30.56	24.65	18.54	76.33

Table 15.3: Influence of different Sterilants on per cent of Browning, Fungal Contamination, Bacterial Contamination and Fresh Explants Survival on Shoot Apical Meristems and Nodal Cuttings of *T. heyneana*

Sterilant Used	*Time of Exposure*	*Browning (Per cent)*	*Fungal Contamination (Per cent)*	*Bacterial Contamination (Per cent)*	*Survival of Explant (Per cent)*
Sodium hypochlorite (1 per cent)	5 min	50.65	77.65	88.43	0
Sodium hypochlorite (1 per cent)	15 min	68.43	78.42	82.56	0
Sodium hypochlorite (1 per cent)	30 min	57.54	55.77	90.77	0
Sodium hypochlorite (1.4 per cent)	5 min	55.56	69.65	78.76	0
Sodium hypochlorite (1.4 per cent)	15 min	45.86	59.76	79.45	0
Sodium hypochlorite (1.4 per cent)	30 min	56.56	62.67	75.54	0
Mercuric chloride (0.1 per cent) + ethyl alcohol (70 per cent)	2 min + 30s	66.67	52.76	62.54	0
Mercuric chloride (0.1 per cent) + ethyl alcohol (70 per cent)	10 min + 30s	68.54	60.65	65.65	0
Mercuric chloride (0.5 per cent) + ethyl alcohol (70 per cent)	5min + 30s	70.56	54.65	68.54	0
Mercuric chloride (0.5 per cent) + ethyl alcohol (70 per cent)	10 min + 1 min	55.32	59.65	70.02	5
Mercuric chloride (1 per cent) + ethyl alcohol (70 per cent)	2 min + 1 min	65.52	69.48	68.52	8
Mercuric chloride (1 per cent) + ethyl alcohol (70 per cent)	5 min + 1 min	55.26	60.28	62.54	10
Mercuric chloride (1 per cent) + ethyl alcohol (70 per cent)	10 min + 1 min	50.01	54.84	62.45	0
Mercuric chloride (2 per cent) + ethyl alcohol (70 per cent)	2 min + 1 min	70.56	62.84	55.48	0
Mercuric chloride (2 per cent) + ethyl alcohol (70 per cent)	5 min + 1 min	79.58	78.25	54.84	0
Mercuric chloride (2 per cent) + ethyl alcohol (70 per cent)	10 min + 1min	80.25	68.35	78.25	0

could not reduce the rate of contamination, the increase in the exposure time resulted in the browning and death of the explants. Antibiotics like streptomycin and chloramphenicol were used to avoid contamination. Chlorine water in different dilutions was also used to further reduce the contamination which didn't fetch fruitful results. Because of the low survival rate of the stem explants, they were not considered for callogenesis.

Initiation, Growth and Maintenance of Callus

Effect of Growth Regulators on Initiation and Growth of Callus

In order to define the optimum concentration of growth regulators essential for rapid initiation and growth of undifferentiated tissues, the MS medium was supplemented with various concentrations of growth hormones (auxins and cytokinins) for the induction of callus from explants like leaf and stem. The medium supplemented with auxin or cytokinin or combinations of auxin with cytokinin showed morphologically different types of callus. After a week of inoculation, callus initiation was observed along the midrib region of leaf explants and cut edges of petiole in *T. heyneana*

Effect of Auxins/Cytokinins Individually on In vitro Callogenesis

The effect of auxins/cytokinins like 2,4–D (0.5-3.0 mg/l), NAA (0.5-3.0 mg/l), Kn (0.1-2.0mg/l) and BAP (0.5-3.0 mg/l) on the initiation of callus from explants in *T. heyneana* is shown in Table 15.4

Table 15.4: Effects of Plant Growth Regulators (PGRs) individually on Callus Induction and Growth from the Explants of *T. heyneana*

Plant Growth Regulators (PGRs)	*Concentration of PGRs (mg/l)*	*Percentage (Per cent) of Callus Induction*	*Response Intensity of Callus*	*Texture of Callus*
Control	–	–	–	
MS +2,4-D	0.5	62	+++	Friable
	1.0	68	+++	Friable
	2.0	70	+++	Friable
	3.0	52	++	Friable
MS + NAA	0.5	52	++	Friable
	1.0	80	++++	Friable
	2.0	68	+++	Friable
	3.0	78	+++	Friable
MS + BAP	0.5	0	–	–
	1.0	0	–	–
	2.0	0	–	–
	3.0	0	–	–

Plant Growth Regulators (PGRs)	*Concentration of PGRs (mg/l)*	*Percentage (Per cent) of Callus Induction*	*Response Intensity of Callus*	*Texture of Callus*
MS + Kn	0.1	20	+	Friable
	0.5	38	+	Friable
	1.0	42	++	Friable
	1.5	57	++	Friable
	2.0	60	+++	Friable

-: No response; +: Poor growth; ++: Moderate growth; +++: Good growth; ++++: Very good growth.

T. heyneana: Medium supplemented with NAA (1mg/l) (Figure 15.2A) alone showed significant amount of friable creamish white callus (80 per cent) (Table 1.5). Medium enriched with 2, 4-D(2mg/l) (Figure 15.2B) induced light green coloured callus (70 per cent). A maximum of 60 per cent callus was induced when the media was enriched with 2 mg/l kinetin (Figure 15.2C) alone which was more or less yellowish in colour and friable in texture. Almost all the concentrations of 2, 4-D, NAA, Kinetin tested produced callus but none of the tried concentrations of BAP induced callus from leaf and petiole explants.

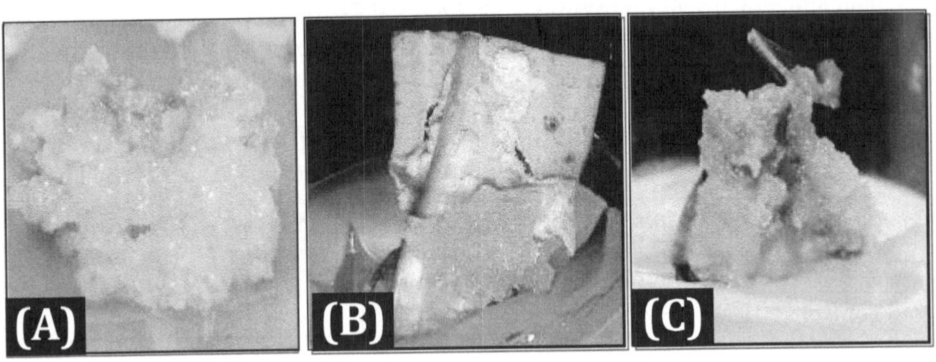

Figure 15.2: Effect of Individual Growth Regulators on the Induction of Callus in *T. heyneana.* (A) NAA; (B) 2, 4-D; (C) Kinetin.

Synergistic Effect of Auxins and Cytokinins

The data on the synergistic effect of combinations of auxins and cytokinins on callus induction from the explants is presented in Table 15.5

T. heyneana: Callus was induced at all the concentrations but the mode of response was different in different combinations (Table 15.6). The leaf explants showed the significant maximum callus induction of 98 per cent and 96 per cent in the medium supplemented with 3mg/l (2, 4 - D) + 2mg/l (Kn) and 4mg/l (2,4-D) + 3mg/l (Kn) respectively. The induced callus was somewhat light green in colour and compact in nature (Figure 15.3A). The response of callus growth was more

Table 15.5: Effects of PGRs in Combination on Callus Induction and Growth from the Explants of *T. heyneana*

PGRs	*Concentration of PGRs (mg/l)*	*Percentage (Per cent) of Callus Induction*	*Response Intensity of Callus*	*Texture of Callus*
MS (Control)	–	–	–	–
MS + 2,4-D+ Kn	0.5 + 0.1	10	+	Compact
	1.0 + 0.5	30	+	Compact
	2.0 + 1.0	48	+ +	Compact
	3.0 + 2.0	98	+ + + +	Compact
	4.0 + 3.0	96	+ + + +	Compact
MS+2,4-D+ BAP	0.5 + 0.1	10	+	Compact
	1.0 + 0.5	15	+	Compact
	2.0 + 1.0	60	+ + +	Compact
	3.0 + 2.0	86	+ + + +	Compact
	4.0 + 3.0	98	+ + + +	Compact
	5.0 + 4.0	98	+ + + +	Compact
MS+NAA+BAP	0.5 + 0.1	10	+	Friable
	1.0 + 0.5	15	+	Friable
	2.0 + 1.0	19	+	Friable
	3.0 + 2.0	20	+	Friable
MS+NAA+Kn	0.5 + 0.1	48	+	Compact
	1.0 + 0.5	37	+	Compact
	2.0 +1.0	28	+	Compact
	3.0 + 2.0	50	+	Compact

–: No response; +: Poor growth; ++: Moderate growth; +++: Good growth; ++++: Very good growth.

vigorous (98 per cent) even in the medium supplemented with 4 mg/l (2,4-D) + 3 mg/l (BAP) and 5 mg/l (2,4-D) + 4 mg/l (BAP) which showed pale green coloured callus (Figure 15.3B) along the midrib region of leaf lamina. The callus induction was increased with the increase in the concentration of growth regulators. Leaf explants showed poor callusing on the media supplemented with NAA and BAP. Maximum callogenic response (20 per cent) was observed in the medium supplemented with 3mg/l (NAA) + 2mg/l (BAP) which showed compact yellow coloured callus (Figure 15.3C). However, in case of media fortified with different concentrations of NAA and Kinetin, callusing was observed in all the tried combinations, whereas the highest response (50 per cent) was observed at 3mg/l (NAA) + 2mg/l (Kn). The obtained callus was compact yellowish green in colour (Figure 15.3D). Poor callusing was seen from petiole explant which was confined to only the cut ends of the explants (Figure 15.3E). Stem explants also showed a poor response (Figure 15.3F) with high rate of bacterial and fungal contamination. Hence, only leaf explants were selected for all further experiments.

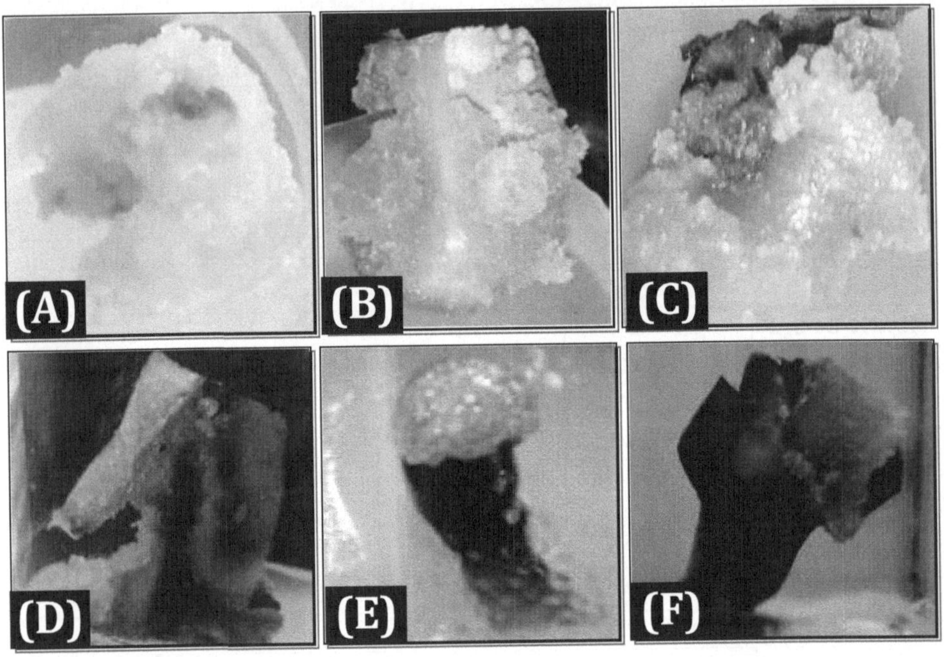

Figure 15.3: Synergistic Effects of Combination of Auxins and Cytokinins on Callus Induction and Growth from the Explants of *T. heyneana*.
(A) 2,4–D + Kn; (B) 2,4-D+BAP; (C) NAA+BAP; (D) NAA+Kn;
(E) Callusing from petiole; (F) Callusing from stem.

Effect of Subculturing and Maintenance of Callus

The effect of subculturing of leaf explants on calli initiated on MS basal medium with various concentrations of growth regulators in *T. heyneana* is shown in Table 15.6

Table 15.6: Effect of Callus Subculturing on Callus Proliferation in *T. heyneana*

Media	*Response of Callus upon Subculturing*	*Figure 15.4*
MS+2,4-D	Callus turned brown in 7 days after subculturing	A
MS+NAA	No callus proliferation on subculturing	B
MS +Kn	Explant turned brown and started to dry up in 2 weeks	C
MS+2,4-D+Kn	Profuse callus proliferation, callus was fresh, healthy and yellowish in colour after 3 weeks of subculturing.	D
MS+2,4-D+BAP	Callus proliferated for a week after subculturing later turned brown.	E

Media	*Response of Callus upon Subculturing*	*Figure 15.4*
MS+NAA+BAP	Callus proliferated after subculturing for 15 days and eventually turned brown due to the accumulation of phenolic compounds.	F
MS+NAA+Kn	Newly generated callus was fresh but the old callus undergone necrosis over time.	G, H

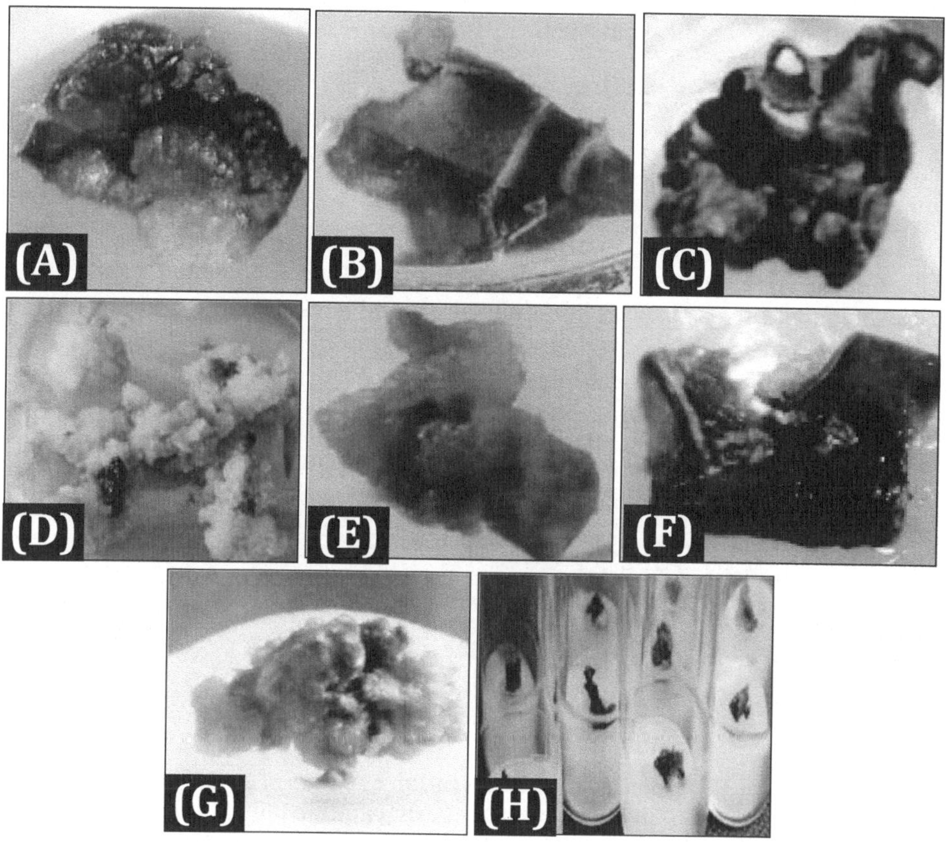

Figure 15.4: Effect of Subculturing on Callus Proliferation in *T. heyneana*. (A) 2,4-D; (B) NAA; (C) Kn; (D) 2,4-D + Kn; (E) 2,4–D + BAP; (F) NAA+BAP; (G) NAA+Kn; (H) Culture tubes showing browning of explants after subculturing.

Figure 15.4 showed the effect of subculturing on the proliferation of callus induced on varying concentrations of growth regulators in *T. heyneana*. In some cases (MS+2,4-D/NAA/Kn/BAP), after subculturing to the fresh medium, the callus turned brown and eventually did not proliferate in both the plants. In a few cases (MS+2,4-D+BAP/NAA+BAP), callus started to proliferate for a week but turned brown due to the accumulation and oxidation of phenolic compounds after 2 weeks of subculturing, eventually leading to cell death. The use of activated charcoal in the medium and pretreating the explant with Vitamin C (an anti oxidant and absorbent) did not diminish the rate of browning in cultures.

The callus obtained on the medium supplemented with 2,4–D + Kn at concentrations of 3+2mg/l showed profuse callusing response on subculturing compared to the remaining combinations of growth regulators, hence this combination was used for mass proliferation of callus in *T. heyneana*. The callus was subcultured after every 30 days and maintained for 4 months before harvesting for further phytochemical analysis.

Microscopic Observation of Callus

Squash preparation of callus tissue showed extremely heterogenous cellular composition ranging from small cells bearing dense cytoplasm to large cells with number of vacuoles in the cytoplasm.

Microscopic examination of callus tissues of *T. heyneana* showed small ovoid (Figure 15.5A), elongated (Figure 15.5B) and sickle shaped cells (Figure 15.C). Non dividing callus cells were generally elongated having a large vacuole (Figure 15.B) while the small actively diving cells were spherical or curved in shape with reduced vacuole and dense cytoplasm (Figure 15.A) forming a mass of aggregated undifferentiated callus by the end of 3 weeks.

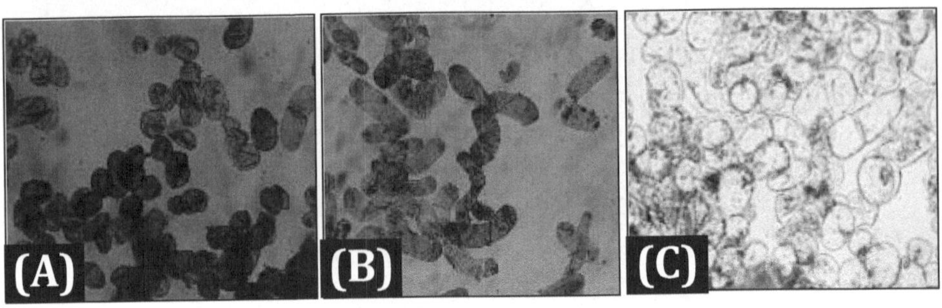

Figure 15.5: Microscopic Observation of Callus of *T. heyneana*.
(A) Small ovoid cells stained using aniline blue (10x);
(B) Normal diving curved callus cells (10X) stained using aniline blue (10x);
(C) Sickle shaped cells ready for division.

Discussion

It is always a big challenge to obtain aseptic cultures free of contamination from the wild plants which are always at a possibly high risk of contamination. More frequently used surface sterilants in tissue culture are NaOCl or CaOCl, mercuric chloride, bromine/chlorine water, ethanol *etc.* However, the use of combination of sterilizing agents is always effective. In the present study the combination of $HgCl_2$ and 70 per cent ethanol showed the highest percentage of aseptic culture in

T. heyneana. Oliveira *et al.* (2003) reported the use of 50 per cent HCl, 95 per cent ethanol and 5 per cent sodium hypochlorite to obtain a maximum percentage of aseptic cultures in *T. fuchsiaefolia.* Mahota *et al.* (2013) obtained highest aspetic cultures in *Holarrhenaanti dysenterica* using 0.2 per cent $HgCl_2$. In *Adenium obesum*, highest explant survival was observed from the treatment of NaOCl and ethanol (Rasad *et al.*, 2015). Similar results were reported in *Catharanthus roseus, Amsonia orientalis* and *Rauwolfia serpentine* (Rashmi and Trivedi2015; Acemi *et al.*, 2012; Bhatt *et al.*, 2008). However, in the current study NaOCl was not effective in preventing contamination of the explants.

In the present study we fail to optimise the surface sterilization technique for *T. heyneana* stem explants. Sixteen different sterilization regimes of treatment were used for treating the explants. Even with the harsh treatment of surface sterilants we could not avoid fungal and bacterial contamination. It was observed that very less percentage of survival of explants were recorded when treated with 0.5 per cent and 1 per cent $HgCl_2$ for the duration of 10 and 5min respectively. Probably, because of the toxic impact of heavy metals, the explants turned black and abstained. The above observations put a platform for the strong conclusion that the plant is harboring endophytes which was also substantiated by Manasa and Nalini (2014) while working on the study of fungal endophytes in *T. heyneana* on the evalution of L-asparaginase activity. As we know that the medium used for culturing tissues are good sources of nutrients which probably influenced the bacterial growth and subsequently led to bacterial contamination of cultures. According to Quambusch *et al.* (2014) the bacterial growth affects the growth rate, increases mortality of cultures, hamper rooting and cause tissue necrosis, resulting in variable growth of cultures finally leading to death of tissues. Mbah and Wakil (2012) opined that the endophytic bacteria cannot be controlled by any of the sterilization technique; the only way to rule out endophytes is the use of antibiotics. Though, attempts have been made to prevent contamination by the use of antibiotics (streptomycin and chloramphenicol), it didn't give any fruitful results in the present study. Use of antibiotics (tetracycline, rifampicin, streptomycin, vancomycin, gentamycin) individually and in combinations could not ultimately control the contamination in *in vitro* cultures of *Dioscorea rotundata* (Mbah and Wakil 2012). Similar observations were also recorded in *Guadua angustifolia* Kunth and sugarcane (Tiwari *et al.*, 2012; Nadha *et al.*, 2012).

In *T. heyneana,* the response of callus induction was observed initially from midrib regions and cut ends of the explants. This result is in line with the results obtained by Das (2010) who reported the development of calli from mid rib as well as cut ends of the explants. Similar reports were given by Verma *et al.* (2016) and Mederos-Molina (2004) while working with Turkish *Crocus* species and *Salvia canariensis.*

The friable creamish white, light green and compact yellow coloured calli was obtained from the leaf explants of *T. heyneana* on the medium tried with different combinations of growth regulators. More vigourous growth of callus was observed on the medium fortified with both auxins and cytokinins. Two combinations of 2,4-D and Kn (1:1 and 1:0.1) initiated callus from the axenic leaf cultures of *T. catharinensis* which is in agreement with the present observation (Pereira *et al.*, 2007). In *Amsonia orientalis,* all the combinations of plant growth regulators successfully induced the callus at various percentages but the most effective combination was 2,4-D and BAP (Acemi *et al.*, 2012). Synergy between the growth regulators on callus induction was confirmed by Bhatt *et al.* (2008) and Rashmi *et al.* (2015) who reported the formation of light green callus in *Rauwolfia serpentina* and *Catharanthus roseaus.*

In the present study, the leaf explants were proved to be better when compared to stem and petiole explants in both the species. Maximum callus induction was observed in leaf explants of *Citrus jambhiri* and Jojoba compared to nodal and root segments (Savita *et al.*, 2011; Kumar *et al.*, 2013). The results are in agreement with the present study. On the contrary, nodal explants in *Solanumtuberosum* and basal explants from *Allium chinense* proved to be better explants for the callus induction (Kumlay *et al.*, 2015).

The browning of callus was observed in the present study upon subculturing to the fresh medium probably, due to the accumulation of phenols in the medium. Therefore, phenolic compounds occurring in these plants play an important role in *in vitro* cultures and regeneration of explants. Phenolic compounds being catalysed by several enzymes *viz.*, polyphenol oxidase (PPO), peroxidase (POD) and phenylanine deaminase (PAL) react with oxygen to form quinine compounds, eventually they undergo dehydration and polymerization reactions which are toxic to callus causing the browning of callus and finally leading to tissue necrosis (Khateeb *et al.*, 2017). Generally, phenolic compounds were concentrated in vacuoles, plastids and cytoplasm. But, when the explants were detached from plants or callus were cut off into fragments during subcuturing, the damaged cells release phenolic compounds into the nutrient medium that hinders the nutrient uptake from the medium. Even, the use of anti oxidants such as ascorbic acid, vitamin C and activated charcoal did not diminish the rate of browning in the present study. Our results are in agreement with the work carried out by Laukkanen *et al.* (1999) who observed the decreased regeneration capacity of callus in Scots pine due to the excessive accumulation of phenolic compounds and lipid peroxidation leading to browning of tissue which could not be controlled by the use of antioxidants. The use of ascorbic acid (567.7 μM) was found to be effective in preventing browning of callus in *Acacia* species which is contrary to the results of the present study (Vengadesan *et al.*, 2002). Squash preparations of callus in *T. heyneana* showed the presence of small ovoid, elongated, crescent shaped cells. Bhagya and Chandrashekar (2010) reported the presence of small round cells, slightly elongated cells and cells showing xylogenesis in *Justicia gendarussa*. Thin walled highly vacuolated cells were observed in *Gardenia latifolia* (Reddy and Saritha 2012).

Conclusion

In this study, for the first time an efficient surface sterilization protocol was developed for *Tabernaemontana heyneana* for the aseptic *in vitro* culturing using leaf explants. A highest percentage (76.33 per cent) of aseptic cultures were obtained by the treatment of 2 per cent bavistin (1 h 30 min) followed by 1 per cent $HgCl_2$ (5min)+70 per cent ethanol (1min) for *T. heyneana* explants. About 98 per cent of callus induction frequencies were obtained in the MS-medium supplemented with 3mg/l 2, 4-D + 2mg/l Kn which is comparatively higher than any other combinations in *T. heyneana*. The protocol for callogenesis and maintenance described in this piece of research work paves the way for forthcoming experiments directing to study the biosynthesis of bioactive metabolites in *T. heyneana*. Moreover, there is a scope for the development of complete *in vitro* protocol for the regeneration of this highly valuable medicinal plant. In addition, the protocol for protoplast isolation and regeneration may be useful in producing somatic hybrid cell lines, with unique composition of pharmaceuticals, through protoplast fusion.

REFERENCES

Acemi A, Ozen F, Kiran R (2012) Development of an efficient callus production protocol for *Amsonia orientalis*: A critically endangered medicinal plant. Eurasia J Biosci 6: 105-112.

Bhagya N, Chandrashekar KR (2010) Effect of auxin concentration of callus induction from *Justicia gendarussa*L. stem and leaf explants. Int J BioSci Tech 3: 27–35.

Bhatt R, Arif M, Gaur AK, Rao PB (2008) *Rauwolfia serpentina*: Protocol optimization for *in vitro* propagation. African J Biotech 7: 4265-4268.

Gamble JS (1958) Flora of the Presidency of Madras. Sri Gouranga Press, India.

Ignacimuthu S, Ayyanar M (2005) Traditional knowledge of Kanitribals in Kouthalai of Tirunelveli hills, Tamil Nadu, India. J Ethnopharmacol 102: 246-255.

Ignacimuthu S, Ayyanar M, Sivaraman K (2006) Ethnobotanical investigations among tribes in madurai district of Tamil Nadu, India. J EthnobiolEthnomed 2: 25-32.

Khateeb WA, Alu'datt M, Haifa Al Zghoul HA, Kanaan R, El-Oqlah A, Lahham J (2017) Enhancement of phenolic compounds production in *in vitro* grown *Rumexcyprius*Murb. ActaPhysiol Plant 39: 1-13.

Kumar S, Mangal M, Dhawan AK, Singh N (2013) Callus induction and plant regeneration from leaf explants of jojoba [*Simmond siachinensis* (Link) Schneider]. Indian J Biotech 12: 544-547.

Kumlay AM, Ercisli S (2015) Callus induction, shoot proliferation and root regeneration of potato (*Solanumtuberosum* L.) stem node and leaf explants under long-day conditions, Biotech Biotech Equip 29: 1075-1084.

Laukkanen H, Haggman H, Kontunen-Scoppia S, Hohtola A (1999) Tissue browning of *In vitro* cultures of Scot pine: Role of peroxidase and polyphenol oxidase. PhysiologicaPlantarum 106: 337-343.

Mahota S, Mehta A, Roy S (2013) Studies on antibacterial effects of bark, seed and callus extracts of *Holarrhenaanti dysenterica* wall. Bioscan 8: 717-721.

Manasa C, Nalini MS (2014) L-asparaginase activity of fungal endophytes from *Tabernaemontana heyneana* Wall. (Apocynaceae), endemic to the Western Ghats (India). IntSch Res Notices 2014: 1-7.

Mbah EI, Wakil SM (2012) Elimination of bacteria from *in vitro* yam tissue cultures using antibiotics. J Plant Path 94: 53-58.

Mederos-Molina S (2004) *In vitro* callus induction and plants from stem and petiole explants of *Salvia canariensis* L. Plant Tissue Cult 14: 167-172.

Murashige T, Skoog F (1962) A revised medium for rapid growth and bioassays with tobacco tissue cultures. Physiol Plant 15: 473-497.

Nadha HK, Salwan R, Kasana RC, Anand M, Sood A (2012) Identification and elimination of bacterial contamination during *in vitro* propagation of *Guadua angustifolia* Kunth. Pharmacogn Mag 8: 93-97.

Naik PM, Al-Khayri JM (2017) Extraction and estimation of secondary metabolites from date palm cell suspension cultures. Methods MolBiol 1637: 319-332.

Oliveira AJB, Carvalho VM, Ferreira A, Sato FY, Machado MFPS (2003) *In vitro* multiplication of *Tabernaemontana fuchsiaefolia* L. (Apocynaceae). R ArvoreVicosa-MG 27: 421-425.

Pereira PS, Ticli FS, Franca SC, Breves CMS, Lourenco MV (2007) Enhanced triterpene production in *Tabernaemontana catharinensis* cell suspension cultures in response to biotic elicitors. Quim nova 30: 1849-1852.

Quambusch M, Pirttila AM, Tejesvi MV, Winkelmann T, Bartsch M (2014) Endophytic bacteria in plant tissue culture: differences between easy and difficult to propagate *Prunusavium* genotypes. *Tree Physiol* 34: 524-533.

Rasad FM, Hasbullah NA, Daud NF, Azis NA, Amin MAM, Lassim MM (2015) Micropropagation of *Adenium obesum* (Dessert Rose) *in vitro*. International conference on agricultural, ecological and medical sciences, Thailand. http: // iicbe.org/upload/8112C0415009.pdf.

Rashmi R, Trivedi MP (2015) Rapid *in vitro* regeneration of an important medicinal and an ornamental plant (*Catharanthusroseus* L). Biochem Anal Biochem 4: 1-6.

Reddy MY, Saritha KV (2012) Callus induction and somatic embryogenesis of *Gardenia latifolia* Ait. Int J CurrSci 4: 83-89.

Roy R, Grover RK, Srivastva S, Kulshreshtha DK (2002) A new stereoisomer of stemmadenine alkaloid from *Tabernaemontana heyneana* Wall. Magn Reson Chem 40: 474-476.

Sathishkumar T, Baskar R, Shanmugam S, Rajasekaran P, Sadasivam S, Manikandan V (2008) Optimization of flavonoids extraction from the leaves of *Tabernaemontana heyneana* Wall. using L16 Orthogonal design. NatrSci 6: 10-21.

Savita, Singh B, Virk GS, Nagpal AK (2011) An efficient plant regeneration protocol from callus cultures of *Citrus jambhiri* Lush Physiol. Mol. Biol. Plants 17: 161-169.

Tiwari S, Arya A, Kumar S (2012) Standardizing sterilization protocol and establishment of callus culture of sugarcane for enhanced plant regeneration *in vitro.* Res J Botany 7: 1-7.

Vengadesan G, Ganapathi A, PremAnand R, Anbazhagan VR (2002) *In vitro* propagation of *Acacia sinuata* (Lour.) Merr. via cotyledonary nodes. Agroforestry Systems 55: 9-15.

Verma SK, Das AK, Cingoz GS, Uslu E, Guel E (2016) Influence of nutrient media on callus induction, somatic embryogenesis and plant regeneration in selected *Turkish Crocus* species. Biotech Reports 10: 66-74.

Wink M, Alfermann AW, Franke R, Wetterauer B, Distl M, Windhovel J (2005). Sustainable bioproduction of phytochemicals by plant *in vitro* cultures: anticancer agents. Plant Genetic Resour. 3: 90-100.

Index

Printed at Repro India Ltd.